D0213737

A Dictionary of Biology

A Dictionary of Biology

M. Abercrombie
C. J. Hickman
M. L. Johnson

AldineTransaction
A Division of Transaction Publishers
New Brunswick (U.S.A.) and London (U.K.)

First paperback printing 2008
Copyright © 1967 by Transaction Publishers, New Brunswick, NJ

All rights reserved under International and Pan-American Copyright Conventions. No part of this book may be reproduced or transmitted in any form or by any means, electronic or mechanical, including photocopy, recording, or any information storage and retrieval system, without prior permission in writing from the publisher. All inquiries should be addressed to AldineTransaction, A Division of Transaction Publishers, Rutgers—The State University, 35 Berrue Circle, Piscataway, New Jersey 08854-8042. www.transactionpub.com

This book is printed on acid-free paper that meets the American National Standard for Permanence of Paper for Printed Library Materials.

Library of Congress Catalog Number: 2008005873
ISBN: 978-0-202-36219-9
Printed in the United States of America

Library of Congress Cataloging-in-Publication Data

Abercrombie, M. (Michael), 1912-1979.
A dictionary of biology / M. Abercrombie, C. J. Hickman, and M. L. Johnson.
p. cm.
Includes bibliographical references and index.
ISBN 978-0-202-36219-9 (alk. paper)
1. Biology—Dictionaries. I. Hickman, C. J. II. Abercrombie, M. L. J. (Minnie Louie Johnson), 1909-1984. III. Title.

QH13.A25 2008
570.3—dc22 2008005873

QH
13
.A25
2008

USING THE DICTIONARY

MANY unfamiliar terms, especially the rarer ones, are defined with the help of other technical terms, perhaps equally unfamiliar. This annoying trick of dictionary-makers could only be avoided by giving a complete account of a large part of biology under each heading, which is impracticable. There is nothing for it but to follow up the terms until you come to an entirely intelligible definition. Every biological technical term used in a definition is, we believe, itself defined elsewhere in the dictionary; though some semi-technical terms, which can be found in any English Dictionary (such as the Pocket Oxford), are omitted.

We put the letters (q.v.) after a term which is defined elsewhere only when its definition adds important information to the subject under discussion. It is not put after every term defined elsewhere. The letters Cf. are intended to point out a contrasting term.

(Bot.) or (Zool.) before a definition means that what follows applies only to plants or to animals respectively. It is not used where the limitation is obvious.

(Adj.) means that the term defined is an adjective, when this is not clear from the definition.

If a word is not defined in the dictionary, try Uvarov's *Dictionary of Science* (Penguin Reference Book R1); and consult it for supplementary information on any chemical term.

AUTHORS' NOTE

THERE are well over a million organisms, each with a name, and tens of thousands of other biological terms. In selecting a mere two thousand for definition we must often have chosen badly. Similarly, we have been forced to limit severely the amount of information we present under each heading, and may well have misplaced the emphasis on many occasions. We should be glad to have criticisms about these points, and about inaccuracies of fact, from users of the dictionary.

Our grateful thanks go to those friends who by their criticism of parts of the work have removed so many mistakes. They must not be blamed for any of the defects which remain. Dr K. W. Dent, Mr D. R. Newth, Miss Pauline Whitby, and Professor J. Z. Young all helped us in this way with the original edition. Many others have greatly assisted us by suggesting emendations for the revised editions, or by generously responding to requests for advice and information. Miss E. R. Turlington has most kindly contributed several of the diagrams. Mrs M. A. Bauer-Nelke's collaboration in the early stages of the work was of great value to us.

M. A.

C. J. H.

M. L. J.

A

ABAXIAL (Dorsal). (Of a leaf surface), facing away from the stem. Cf. *Adaxial.*

ABDOMEN. In vertebrates: region of the body containing the viscera other than heart and lungs (i.e. intestine, liver kidneys, etc.); in mammals, but not in other vertebrates, bounded anteriorly by diaphragm. In arthropods: posterior group of segments similar to each other.

ABDUCENS NERVE. Sixth cranial nerve of vertebrates. Almost entirely motor, supplying external rectus eye-muscle. See *Eye-muscles.* A ventral root.

ABSCISSION LAYER. Layer at base of leaf-stalk in woody dicotyledons and gymnosperms, in which the parenchyma cells become separated from one another through dissolution of the middle lamella before leaf-fall.

ABYSSAL. Inhabiting deep water (roughly below 1,000 metres).

ACANTHODII. Group of fossil fish (mainly Devonian, 350 to 400 million years ago, but lasted through 150 million years to Permian). An order of Placodermi. Notable because many of them had not only pectoral and pelvic fins, but a row of other paired fins between them, each bearing a prominent spine along its anterior margin.

ACARINA (ACARIDA). Order of Arachnida including mites and ticks; some are important parasites, e.g. *Sarcoptes* causing scabies, cattle tick carrying redwater fever.

ACCESSORY NERVE. Eleventh cranial nerve of tetrapod vertebrates. Really a branch of the vagus, clearly separate only in mammals.

ACCOMMODATION Changing the focus of the eye. In man and a few other mammals occurs by changing curvature of lens; at rest, lens is focused for distant objects; it is focused for near objects by becoming more convex with the contraction of the ciliary muscles in ciliary body q.v. (see Fig. 3 p. 86). Few mammals can accommodate. Most birds and reptiles accommodate by changing curvature of the lens; in fish and amphibians, lens is moved backwards and forwards in relation to retina (as in focusing a camera).

ACETABULUM. Cup-like hollow on each side of hip girdle into which head of femur (thigh bone) fits, forming hip joint, in tetrapod vertebrates.

ACETYLCHOLINE (ACh). Substance secreted at the ends of many nerve fibres (cholinergic fibres) when nerve impulses arrive there. Where a nerve fibre ends at a synapse, ACh may be the agent which stimulates the contiguous nerve cell and hence in effect 'passes the impulse on'; and similarly where the fibre connects with an effector. After secretion ACh is very rapidly destroyed by the enzyme cholinesterase. It also takes part in the mechanism of conduction along a nerve fibre. It is the acetyl ester of choline (q.v.).

ACHENE. Dry, one-seeded fruit formed from a single carpel, with no special method of opening to liberate the seed; may be smooth-walled, e.g. buttercup; feathery, e.g. traveller's joy; spiny, e.g. corn buttercup; or winged (*samara*), e.g. sycamore.

ACHLAMYDEOUS. (Of flowers), lacking petals and sepals, e.g. willow.

ACID DYES. Dyes consisting of an acidic organic grouping of atoms (anion) which is the actively staining part, combined with a metal. Stain particularly cytoplasm and collagen. Cf. *Basic dye.*

ACOELOMATE. Having no coelom (e.g. the phyla Coelenterata, Platyhelminthes, Nemertea, Nematoda).

ACOUSTIC. Concerned with hearing. *A. nerve*, Auditory nerve (q.v.).

ACQUIRED CHARACTERISTICS, INHERITANCE OF. Transmission to offspring of variations, which appeared in the parents as responses to environmental influences. E.g. exposure to sunlight causes darkening of the skin of white human beings, compared to others less exposed: an acquired characteristic. If this acquired characteristic were inherited, offspring of the darker parents would then tend, even if only very slightly, to be darker than offspring of the lighter parents, when both groups of offspring are reared in equal sunlight. The view that such inheritance occurs is commonly known as Lamarckism (q.v.) or Neo-Lamarckism. It is not widely thought that such inheritance is of any importance in organisms reproducing sexually. The gametes are not affected by these acquired variations in such a way as to transmit them to the offspring. It has however been recently shown that natural selection (q.v.) may change successive generations of a population so that a characteristic at first acquired only in response to the environment may come to develop independently of the environmental stimulus (*genetic assimilation*). Organisms reproducing asexually may, of course, hand on to their offspring part of the body complete with its acquired characteristics.

ACRANIA (CEPHALOCHORDATA). Sub-phylum of Chordata containing only species of amphioxus, all marine. Unlike Vertebrata they have no brain, skull, or cartilaginous or bony skeleton; but they have typical chordate dorsal tubular nerve-cord with double nerve roots, notochord, gill-slits, muscle blocks (myotomes); and they have an unexpected feature, found in no other chordate though in many invertebrates – nephridia as excretory organs. They may be closely related to early ancestors of fish and other vertebrate groups. A very early fossil probably belonging to the group has been discovered (*Jaymoytius* from the Silurian).

ACROPETAL. (Bot.). Development of organs in succession towards apex, the oldest at base, youngest at tip, e.g. leaves on a shoot. Cf. *Basipetal.*

ACROSOME. Part of head of animal sperm, usually forming a

cap over the nucleus. Function in fertilization not yet known.

ACTH (CORTICOTROPIN). Adreno-cortico-tropic hormone, secreted by anterior lobe of pituitary, controlling secretory activity of adrenal cortex. A polypeptide.

ACTINOMORPHIC. (Of flowers), regular; capable of bisection vertically in two or more planes into similar halves, e.g. buttercup. Also known as *radially symmetrical*, a term used to describe animals having a similar organization, e.g. jelly-fish. Cf. *Zygomorphic*.

ACTINOMYCETE. Member of genus *Actinomyces*, group of filamentous bacteria parasitic in mammals. Is also widely used to include all filamentous bacteria.

ACTINOPTERYGII. A class of fish (or sometimes regarded as a subclass of the class Osteichthyes); includes all common fish except sharks and skates. Characterized by bony skeleton; absence of a central skeletal axis in paired fins, their skeletal support being like the ribs of a fan; no opening of nostrils into mouth. Ganoid scales (q.v.) in primitive species. Cf. *Chondrichthyes* and *Choanichthyes*, the other two classes of living fish. Appear first in Devonian (350 to 400 million years ago); originally fresh-water, but later colonized the sea.

ACTINOZOA (ANTHOZOA). Sea-anemones, corals, sea-pens, etc. Class of Coelenterata (of subphylum Cnidaria). No medusa stage; polyp more complexly organized than that of other Coelenterata; possessing an intucking of ectoderm into coelenteron (stomodaeum); and vertical partitions in coelenteron (mesenteries). See *Alcyonaria*. Cf. *Hydrozoa*, *Scyphozoa*.

ACTION POTENTIAL. Of a nerve impulse; a localized change of electrical potential between the inside and outside of a nerve fibre, which marks the position of an impulse as it travels along the fibre. In the absence of an impulse the inside is electrically negative to the outside (the resting potential); and during the passage of an impulse past any point on the fibre it changes momentarily to positive. This wave of potential change is the most easily detectable and measurable aspect of an impulse. A similar action potential occurs in a muscle fibre when it is stimulated.

ACTIVATED SLUDGE. Material consisting largely of Bacteria and Protozoa, used in, and produced by, one method of sewage disposal. Sewage is mixed with some activated sludge and agitated with air; organisms of the sludge multiply and purify the sewage, and when it is allowed to settle they separate out as a greatly increased amount of activated sludge. Part of this is added to new sewage and part disposed of.

ACTIVE TRANSPORT. Transfer of substance from region of low to region of high concentration, accomplished by means of expenditure of energy from metabolism. Probably all cells are capable of performing this.

ACTOMYOSIN. A complex of two proteins, actin and myosin, forming a major constituent of muscle. Shortening of actomyosin fibrils produces contraction of muscles. Myosin is very closely associated with an enzyme whose activity liberates energy for this contraction from ATP (q.v.).

ACUSTICO-LATERALIS SYSTEM. Lateral line system (q.v.).

ADAPTATION. (1) Evolutionary. Any characteristic of living organisms which, in the environment they inhabit, improves their chances of survival and ultimately of leaving descendants, in comparison with the chances of similar organisms without the characteristic; natural selection therefore tends to establish adaptations in a population. An adaptation *to* a particular feature of the environment means a characteristic which is an adaptation because it reduces destruction by that particular feature. An adaptation *to* a particular *activity* of an organism (e.g. to flying) means simply a characteristic which makes possible or improves performance of that activity without necessarily being measured in terms of survival, though usually that is implied. (2) Physiological. Change in an organism as a result of exposure to certain environmental conditions which makes it react more effectively to these conditions. (3) Sensory. Change in excitability of a sense-organ as a result of continuous stimulation such that a more intense stimulus becomes necessary to produce the same activity, e.g. contact of an object with the skin at once excites the touch receptors; but if contact is maintained the touch receptors quickly cease to respond, though they will respond to an increased stimulus. Different receptors differ much in the extent of their adaptation.

ADAPTIVE (INDUCIBLE) ENZYME. Enzyme formed by an organism in appreciable amounts only in response to the presence of its substrate or a similar substance. Bacteria especially are known to adjust their enzyme make-up in this way.

ADAPTIVE RADIATION. Evolution, from a primitive type of organism, of several divergent forms adapted to distinct modes of life. E.g. at beginning of Tertiary the basal stock of placental mammals radiated into many forms adapted to running, flying, swimming, burrowing, etc.

ADAXIAL. (Of a leaf surface), facing the stem. Cf. *Abaxial.*

ADENOSINE TRIPHOSPHATE. ATP (q.v.).

ADIPOSE TISSUE. Fatty tissue. Connective tissue, the cells of which contain large globules of fat.

ADRENAL (SUPRARENAL) GLAND. An organ of hormone secretion in vertebrates. There is a single pair, one near each kidney, in man and other mammals; but there are multiple adrenals in many other vertebrates. In all tetrapods each gland has two components, distinct in function but closely fused together. (*a*) Medulla, the inner part of the gland in mammals, embryologically derived from nervous tissue (probably neural crest), secreting adrenalin. Medullary tissue seems to have largely an emergency function, secreting adrenalin when the animal is driven to fight or flee; it is not essential for a quiet life. (*b*) Cortex, the outer part of the gland in mammals, embryologically derived from the lining of the coelom, secreting various hormones chemically related to sex-hormones (steroids). These hormones control to varying degrees the salt and water balance of the body, carbohydrate metabolism and other processes; and they are important in ensuring an effective adaptation to stress, such as injury or violent exertion. Cortical hormone secretion is controlled by the pituitary (see *ACTH*). The adrenal cortex is indispensable for life. Medullary and cortical tissues are separated into distinct organs in fishes.

ADRENALIN (ADRENIN, EPINEPHRINE.) Hormone secreted by medulla of adrenal gland (q.v.). Substances with similar action (mainly noradrenalin) are secreted at many nerve endings of sympathetic nervous system, and this accounts for similarity of the action of adrenal medullary hormone to the effects of massive stimulation of sympathetic system (increased heart rate, blood pressure and blood-sugar; dilation of blood-vessels of muscles and contraction of those of skin and viscera; widening of pupil; erection of hair, etc.). Adrenalin can be chemically synthesized; it is amino-hydroxyphenyl-propionic acid.

ADRENERGIC. Of a motor nerve fibre, secreting at its end substances especially noradrenalin (q.v.), similar to adrenalin (q.v.) when nerve-impulse arrives there. These substances stimulate the effector innervated by the nerve fibre. Many vertebrate sympathetic motor nerve fibres are adrenergic. Cf. *Cholinergic.*

ADVENTITIOUS. Arising in abnormal position; of roots, developing from part of plant other than roots, e.g. from stem or leaf cutting; of buds, developing from part of plant other than in axil of leaf, e.g. from root.

AERENCHYMA. Tissue of thin-walled cells with large, air-filled intercellular spaces, found in roots and stems of some aquatic and marsh plants.

AEROBIC RESPIRATION. Respiration (q.v.) in presence of free (i.e. gaseous or dissolved) oxygen. Cf. *Anaerobic respiration.*

AESTIVATION. (1) (Bot.). The arrangement of the parts in a flower-bud. (2) (Zool.). Dormancy during summer or dry season; it occurs e.g., in lung-fish (Dipnoi). See *Hibernation.*

AFFERENT. Leading towards, e.g. of arteries leading to vertebrate gills; or of nerve-fibres conducting impulses towards central nervous system (sensory fibres). Cf. *Efferent.*

AFTER-RIPENING. Refers to dormancy (see *Dormant*) exhibited by certain seeds, e.g. hawthorn, apple, which, although embryo is apparently fully developed, will not germinate immediately seed is formed. Embryo will not grow even when removed from seed coat and provided with favourable conditions but has to undergo certain chemical and physical changes before it is capable of growth.

AGAMOSPERMY. All types of apomixis (q.v.) in which embryos and seeds are formed by asexual means; excludes vegetative reproduction.

AGAR (AGAR-AGAR). Mucilage (polysaccharide) obtained from certain seaweeds; forms a gel with water and is used to solidify culture media on which micro-organisms are grown.

AGGLUTINATION. Sticking together, e.g. of bacteria (one of the effects of antibodies); or of red blood corpuscles (when blood of incompatible blood groups is mixed).

AGNATHA. Class of vertebrates (sometimes made a sub-phylum, the other vertebrate classes then being grouped as the sub-phylum Gnathostomata). Represented now by very few species (order Cyclostomata, i.e. the lampreys and hagfishes). Aquatic, fish-like in many respects, but without jaws, and the two pairs of fins or legs characterizing nearly all other vertebrates are absent (though there may be one pair). The earliest fossil vertebrates known belong to this group; these are the Ostracoderms, which lived in the Silurian (possibly Ordovician) 400 to 450 million years ago, and were probably ancestral to all other vertebrates.

AIR-BLADDER. Swim-bladder (q.v.).

AIR-SACS. Of birds: thin-walled, air-filled extensions of the lungs, lying in abdomen and thorax, and extending even into some of the bones. Of some insects: thin-walled diverticula of tracheae. In both, important in respiration.

ALBINISM. Failure of development of skin pigments. In mammals, including man, commonly due to a single recessive gene.

ALBUMEN. Egg-white of birds and some reptiles. A solution of protein in water, between the ovum (the yolk) and the shell membranes. Secreted by the oviduct. It is eventually absorbed by the embryo.

ALCYONARIA. Soft corals, sea pens, etc.; an order of Coelenterates, class Actinozoa. Have eight tentacles and eight mesenteries; unlike the ordinary corals and the anemones which commonly have many.

ALEURONE GRAINS. Granules of protein occurring in storage regions of plants; common in seeds.

ALGAE. Seaweeds, etc. Sub-division of Thallophyta. Unicellular plants; or multicellular, with a filamentous flattened, ribbon-like thallus; with relatively complex internal organization in higher forms. Distinguished from other thallophytes (Fungi) by presence of chlorophyll. Asexual reproduction by fragmentation of thallus, by zoospores, or by non-motile spores. Sexual reproduction isogamous or anisogamous, both gametes motile, or male alone, or both non-motile. Aquatic plants or plants of damp situations, e.g. seaweeds, those forming green scums on ponds, green stains on damp, shaded walls, tree trunks, etc. Including the following classes: Cyanophyceae (Myxophyceae) (blue-green algae); Chlorophyceae (green algae); Bacillariophyceae (diatoms); Phaeophyceae (brown algae); Rhodophyceae (red algae); the Chrysophyceae, Dinophyceae, Xanthophyceae and others are mostly unicellular, planktonic.

ALIMENTARY (ENTERIC) CANAL. The gut; a tube concerned with digestion and absorption of food. In some animals it has one opening only (Coelenterates, flatworms), but in most it has an opening (mouth) into which food is taken and another (anus) from which unassimilated material is ejected.

ALKALOIDS. Group of nitrogen-containing, basic organic compounds present in plants of a few families of Dicotyledons, e.g. Solanaceae, Papaveraceae; possibly end-products of nitrogen metabolism. Of great importance because of their poisonous and medicinal properties, e.g. atropine, cocaine, morphine, nicotine, quinine, strychnine.

ALLANTOIS. Sac-like outgrowth of ventral side of hinder part of gut present in embryos of amniote vertebrates; represents a large and precocious development of urinary bladder. Allantois grows during development so that it extends right outside the embryo proper, to lie in wall of yolk-sac of birds and reptiles, or under chorion of mammals. It is always covered with connective tissue containing a rich network of blood-vessels, communicating with embryonic circulation. In reptiles and birds, respiration takes place via these blood-vessels, which lie immediately under outer layer of yolk-sac which itself is pressed close against inside of shell; excretory products are stored in allantoic cavity (see *Uricotelic*); and the greater part of allantois is left behind in the shell at hatching. In placental mammals the allantoic blood-vessels supply blood to placenta (q.v.) serving not only for respiration but for nutrition and excretion; cavity may be large and accumulate urine, but is often very small; most of allantois and its blood-vessels are detached from embryo at birth. See Fig. 8, p. 184.

ALLELES (ALLELOMORPHS). Two or more genes (q.v.) are said to be alleles (of each other) or allelomorphic (to each other) when they (1) occupy the same relative position (locus) on homologous chromosomes, and, when in the same cell, undergo pairing during meiosis (q.v.); (2) produce different effects on the same set of developmental processes; and (3) can mutate one to another. Several genes allelomorphic to each other are called an allelomorphic series; not more than two members of a series can simultaneously be present in a normal diploid cell.

ALLOGAMY. (Bot.). Cross fertilization.

ALLOPATRIC. (Of geographical relationship of different species or sub-species), not occurring together, i.e. having different areas of distribution. Cf. *Sympatric*.

ALLOPOLYPLOID. A polyploid (q.v.) organism to which two different species have each contributed one or more sets of chromosomes. Cultivated wheats are probably allopolyploids. See *Allotetraploid*. Cf. *Autopolyploid*.

ALL-OR-NONE LAW. Statement about certain irritable tissues, that they have in standardized conditions only two possible reactions to stimuli of whatever intensity; either no response, or response of one invariable strength. Applies to nerve-cells; and, with qualification that rapidly *repeated* stimuli cause stronger contraction, to muscle-fibres.

ALLOTETRAPLOID (AMPHIDIPLOID). Allopolyploid (q.v.) which arises when an ordinary hybrid between two different species, containing a set of chromosomes from each parent, doubles its chromosome number. An ordinary hybrid is usually sterile because its chromosomes cannot pair during meiosis. But if it becomes an allotetraploid it solves this difficulty because each chromosome then has a homologue with which it can pair. An entirely new species is thus immediately created. Tobacco probably originated in this way. Since interspecific hybridization and successful polyploidy (q.v.) is rarer in animals than in plants, allotetraploidy is known at present only in plants. Artificial preparation of allotetraploids has great scope in the production of new agricultural and horticultural varieties. Doubling of the chromosome number can be achieved in several ways, particularly by colchicine (q.v.).

ALTERNATION OF GENERATIONS. In life cycle (q.v.), alternation of a generation having sexual reproduction with a generation having asexual reproduction. The sexually and asexually reproducing forms are often very different from each other. Occurs among animals in, e.g. hydroids, jelly fish, tapeworms; in these both generations are diploid. In plants seen most clearly in, e.g., ferns, where the two generations are independent. The fern plant is a diploid *sporophyte* and reproduces asexually by formation of haploid spores following meiosis. Germination of the spores initiates the *gametophyte* generation, a small prothallus (q.v.), which reproduces sexually. Male and female gametes fuse together to form a zygote which develops into a new fern plant.

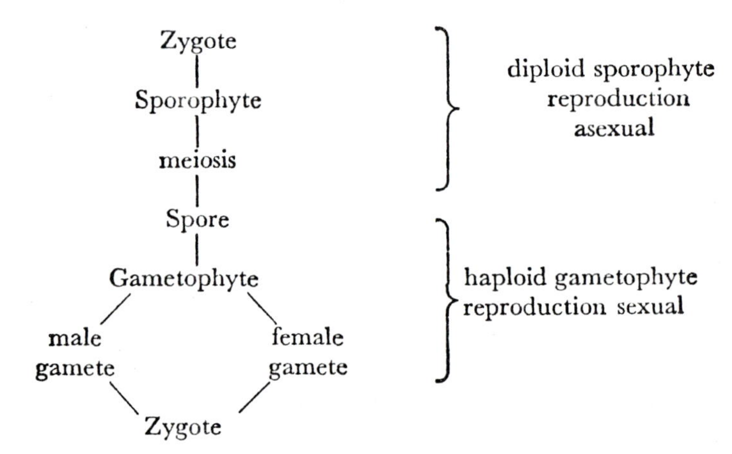

Great differences exist between the plant groups with respect to the relative prominence and degree of independence shown by gametophyte and sporophyte generations. In many members of the Thallophyta, spores are not produced by the diploid generation, which cannot therefore be termed 'sporophyte'. Nevertheless, one can recognize, as in all sexually reproducing plants, an alternation between haploid and diploid phases in the life history. In the majority of Phycomycete and Ascomycete fungi the mycelium is haploid and commonly gives rise to asexually produced haploid spores as well as to gametes. The diploid phase is confined to the zygote, meiosis occurring with the first division of the zygote nucleus. In the Basidiomycetes on the other hand, the mycelium consists of dikaryon (q.v.) cells, a condition usually considered equivalent to the diploid condition. The haploid phase, initiated by the basidiospore, is brought to an end early in the life cycle by union of two basidiospores, or anastomosis of hyphae produced by them, to form a dikaryon. In the Algae, various conditions exist, e.g. dominant gametophyte with diploid phase confined to the zygote; dominant diploid plant bearing dependent gametophyte; distinct gametophyte and true sporophyte plants. In mosses and liverworts the dominant generation, the moss or liverwort plant, is the gametophyte; the sporophyte, the capsule, is small, nutritionally dependent on the gametophyte. In flowering plants, there are separate male and female gametophytes reduced to microscopic proportions. The male gametophyte is shed as the pollen grain; the female gametophyte, the embryo sac, is retained on the sporophyte in the ovule. The sporophyte generation is the plant itself (herb, shrub, tree).

ALVEOLUS. (1) Minute air-filled sac in vertebrate lung, thin-walled and surrounded by blood-vessels. There are large numbers of alveoli in each lung, and it is through their surfaces that the respiratory exchange of oxygen and carbon dioxide occurs. In most vertebrates they connect with the mouth by a system of ramifying air-tubes (bronchi and bronchioles). (2) Expanded sac of secretory cells which forms internal termination of each duct in many glands (e.g. mammary glands). (3) Cavity in jaw-bone into which a tooth fits.

AMETABOLA. Apterygota (q.v.).

AMINO ACID. Organic compound containing both basic amino (NH_2) and acidic carboxyl (COOH) groups. Fundamental constituents of living matter because some hundreds or thousands of amino acid molecules are combined to make each protein molecule. There are twenty different amino-acids commonly found in proteins (q.v.), and there are a few other rare ones. Essential formula of naturally occurring ones (α-amino-acids) is $R\text{-}CH(NH_2)\text{-}COOH$, where R is a variable grouping of atoms (fundamentally a carbon chain or ring), an amino group always being attached to the carbon atom next to the carboxyl group. Amino acids are synthesized by autotrophic organisms such as most green plants. Certain ('essential') amino acids must, like vitamins (q.v.), be obtained from the environment by heterotrophic organisms. There are eight such for man (valine, leucine, phenylalanine, tryptophane, lysine, isoleucine, methionine, threonine), and almost the same list is known for a ciliate, an insect, a bird, and the rat. Other ('non-essential') amino acids are not necessarily required from the environment, since the organism can synthesize them, though in some cases only from essential ones.

AMITOSIS. Uncommon process of division of nucleus by simple constriction into two halves, without formation of a spindle, dissolution of nuclear membrane, or appearance of chromosomes, all of which occur in mitosis (q.v.). Formation of daughter nuclei with identical sets of chromosomes, as after mitosis, almost certainly does not result. Whether duplication of chromosomes occurs and the diploid numbers of chromosomes is approximately maintained in nuclei originating by amitosis is uncertain. Occurs, e.g. in endosperm tissue of flowering plants, macronucleus of Ciliophora, and mammalian cartilage.

AMMOCOETE. Larva of lamprey (a cyclostome). See *Endostyle*.

AMMONITE. One of a group of extinct cephalopod molluscs, related to the living pearly nautilus, with chambered shell. Abundant in the Mesozoic.

AMNION. The embryo of amniote vertebrates (reptiles, birds, mammals) develops in a fluid-filled sac. The wall of this sac has two layers of epithelium with mesoderm and coelomic space between, formed usually by folds of extraembryonic ectoderm and mesoderm which grow up around, and eventually roof over, the embryo. The inner epithelium of the wall is the amnion, though the term is also applied to the whole sac. The outer epithelium is usually called the chorion. Amniotic fluid within the sac provides a fluid environment for the embryo, necessary for animals reproducing on land. In mammals the fluid probably cushions embryo against distortion by maternal organs pressing on it. See Fig. 8, p. 184.

AMNIOTE. Any reptile, bird, or mammal. The *Amniota* form a grouping of vertebrate classes contrasted with the Anamniota. Consists of the essentially land-living vertebrate classes whose embryos have an amnion and allantois (Reptilia, Aves, Mammalia).

AMOEBA. Genus of rhizopod Protozoa. Single-celled animals of irregular and constantly changing shape and semi-fluid consistency, moving and feeding by projection of temporary processes (pseudopodia) from their surfaces. Sometimes erroneously regarded as the most primitive animals and ancestral to others.

AMOEBOCYTE. Cell capable of active amoeboid movement found in blood and other body fluids of invertebrates. Often phagocytic.

AMOEBOID. Moving by pseudopodia (q.v.).

AMPHIBIA. A class of vertebrates. Represented now by three orders:

frogs, toads (Anura); newts, salamanders (Urodela); tropical burrowing worm-like Apoda. In the course of evolution they were the first vertebrates to inhabit the land (late Devonian, about 370 million years ago), being descended immediately from fish (Choanichthyes). The early Amphibia were immediate ancestors of the reptiles which themselves gave rise to mammals and birds. Amphibia differ from fish in having the four pentadactyle (q.v.) legs typical of tetrapods, hip girdle jointed to the vertebral column at the sacrum, and an ear-drum connected to the inner ear by a rod of bone (columella auris), which crosses the middle-ear. They differ from reptiles in that fertilization is not accomplished by coition, and the eggs are unprotected by a shell and embryonic membranes. Consequently most Amphibia have to become temporarily aquatic for the purpose of reproduction. Fossil Amphibia are distinguished from fossil reptiles by, e.g. having a single vertebra concerned in the sacrum, instead of two. Modern Amphibia have diverged far from those which were ancestral to Reptilia, losing much of their bony skeleton.

AMPHICRIBRAL BUNDLE. See *Vascular Bundle.*

AMPHIDIPLOID. Allotetraploid (q.v.).

AMPHIMIXIS. True sexual reproduction (q.v.) as opposed to *Apomixis* (q.v.).

AMPHINEURA. Class of Mollusca, including chitons; small group of marine animals with some primitive features.

AMPHIOXUS. Primitive chordate animal of genus *Branchiostoma* or *Asymmetron,* up to two inches long, found in localized areas of shallow seas in many parts of the world. See *Acrania.*

AMPHIPODA. Order of Crustacea, including freshwater shrimps.

AMPHIVASAL BUNDLE. See *Vascular Bundle.*

AMYLASES (DIASTASES). Group of enzymes which split starch or glycogen to maltose. Widely distributed in plants and animals, e.g. in malt, pancreatic juice.

ANABOLISM. Synthesis by living things of complex molecules from simpler ones. See *Catabolism, Metabolism.*

ANAEROBIC. Living in absence of free oxygen (gaseous or dissolved). *Anaerobic respiration*: liberation of energy by breakdown of substances not involving consumption of oxygen. See *Respiration, Fermentation.* Cf. *Aerobic.*

ANALOGOUS. (1) An organ of one species is said to be analogous to an organ of another when both organs have the same function (q.v.) and when they are not homologous (q.v.); e.g. tendrils of pea and vine, or eyes of squid and vertebrates, are analogous. Possession of analogous organs does not imply a close evolutionary relationship of the organisms bearing them. It merely indicates adaptation to similar conditions. (2) May also be used simply for similarity of function, whether or not the organs in question are homologous.

ANAMNIOTA. Grouping of vertebrate classes sometimes used in classification in contrast to Amniota (q.v.). Consists of Agnatha, fishes, and Amphibia.

ANANDROUS. (Of flowers) lacking stamens.

ANAPHASE. Stage of mitosis (q.v.) or meiosis (q.v.) when daughter chromosomes are separating towards poles of spindle.

ANATROPOUS. (Of ovule) inverted through 180°, micropyle pointing towards placenta. C.f. *Orthotropous*, *Campylotropous*.

ANDRODIOECIOUS. Having male and hermaphrodite flowers on separate plants.

ANDROECIUM. Collective name for the stamens of a flower (q.v.).

ANDROGEN. General name for any substance with male sex hormone activity in vertebrates, i.e. responsible for development and maintenance of many male sexual characteristics. Androgens are tested by the large growth of comb of castrated cock which they produce. Natural androgens in vertebrates are steroids, produced mainly by testis, to small extent by ovary and adrenal cortex. See *Testosterone*, *Oestrogen*.

ANDROMONOECIOUS. Having male and hermaphrodite flowers on the same plant.

ANEMOPHILY. Pollination by wind. Cf. *Entomophily*.

ANEUPLOID. Having more or less than an integral multiple of the haploid number of chromosomes; therefore genetically unbalanced Cf. *Euploid*.

ANGIOSPERMAE. Flowering plants; sub-division of Spermatophyta. Distinguished from other sub-division, Gymnospermae, by having the ovules borne within a closed cavity, the *ovary*, formed by the megasporophyll; after fertilization the ovary becomes a *fruit*, enclosing one or more seeds. Micro- and mega-sporophylls (stamens and carpels) borne in flowers. Gametophyte generations very reduced. Female gametophyte develops entirely within wall of megaspore which at maturity is a large cell containing eight nuclei, the embryo sac (q.v.). Male gametophyte, initiated by pollen grain (microspore), consists of two non-motile male gametes and a tube cell, within pollen tube. Xylem characteristically has vessels. Includes two classes, Monocotyledoneae and Dicotyledoneae, distinguished by the number of seed leaves (cotyledons) in the embryo, one in the former, two in the latter class.

ANGSTROM (Å). See *Micron*.

ANIMAL POLE. (Zool.). Point on surface of egg nearest to its nucleus, marking one end of the graded distribution of substances which occurs in most eggs. The other end, at opposite side of egg, is the *vegetal pole*. The *animal-vegetal* axis between the two poles passes through the nucleus. Quantity of yolk, when present, is usually graded along the axis, and is least at animal, most at vegetal pole.

ANISOGAMY. Condition in which gametes are unlike; may be (1) restricted to condition in which they differ in size, but are similar in form (excluding therefore oogamy, q.v.); or may

(2) include oogamy, and all other grades of difference. Cf. *Isogamy.*

ANNELIDA (ANNULATA). Ringed or segmented worms (bristle-worms, earthworms, leeches). Phylum of animals having well-marked metameric segmentation; coelom; blood system; nephridia; well-defined nervous system. They are thus more complexly organized than flatworms and roundworms. Includes classes Chaetopoda, Archiannelida, Hirudinea, Gephyrea.

ANNUAL. Plant that completes its life-cycle, from seed germination to seed production, followed by death, within a single season. Cf. *Biennial*, *Ephemeral*, *Perennial.*

ANNUAL RING. Annual increment of secondary wood (xylem) in stems and roots of woody plants of temperate climates. Because of the sharp contrast in size between small wood elements formed in the late summer and large elements formed in spring the limits of successive annual rings appear in a cross section of stem as a series of concentric lines.

ANNULAR THICKENING. Internal thickening of wall of a xylem vessel or tracheid, in form of rings at intervals along its length. Occurs in cells of protoxylem and, whilst providing mechanical support, permits longitudinal stretching as neighbouring cells grow.

ANNULATA. Annelida (q.v.).

ANNULUS. (1) Ring of tissue surrounding the stalk (stipe) of fruit bodies of certain basidiomycete Fungi, e.g. mushroom; (2) Line of cells concerned in the opening of moss capsules and fern sporangia to liberate spores.

ANOPLURA. Sucking lice. Order of small exopterygote insects, parasitic on mammals, whose blood they suck. Includes human louse, important as the carrier of typhus.

ANOXIA. Deficiency of oxygen in the tissues.

ANTAGONISM. (1) Interference with, or inhibition of, growth of one kind of organism by another through the creation of unfavourable conditions, e.g. by exhaustion of food supply or by production of a specific antibiotic substance (e.g. penicillin). (2) Of drugs, hormones, etc., producing opposing effects (cf. *Synergism*). Includes *ion antagonism*, the prevention of effects of individual ions by other ions present. E.g. magnesium is an anaesthetic, calcium opposes this action. See also *Physiological saline.* (3) Of muscles, producing opposite movements so that contraction of one must be accompanied by relaxation of the other (and reflexes usually assure this).

ANTENNA. First appendage (paired) of head of Insecta and Myriapoda, second of Crustacea; usually much jointed, whip-like, mobile; function sensory (touch, smell, etc.) but in some Crustacea may be used for swimming or attachment.

ANTENNULE. First appendage (paired) of head of Crustacea, usually sensory.

ANTERIOR. (1) Of lateral flowers, that part farthest away from the main axis, i.e. facing the bract. See Fig. 4, p. 91. (2) Of animals situated at or relatively nearer to, the front (head) and (usually the end directed forward when the animal is moving). In human anatomy, anterior side is front surface, which is equivalent to *ventral* side of other Mammals.

ANTERIOR ROOT. Of nerve, synonymous with ventral root (q.v.).

ANTHELMINTHICS. Drugs used to remove parastic worms (helminths) from their hosts.

ANTHER. Terminal portion of a stamen (q.v.), containing pollen in *pollen sacs*.

ANTHERIDIUM. Male sex organ of Algae, Fungi, Bryophyta and Pteridophyta.

ANTHEROZOID. Synonymous with *Spermatozoid* (q.v.).

ANTHESIS. Flowering.

ANTHOCYANINS. Red, violet, and blue pigments (glycosides, q.v.) occurring in solution in cell-sap of flowers, fruits, stems, and leaves. Responsible for some autumn colourings of leaves and tinting of young shoots and buds in spring.

ANTHOXANTHINS. Water soluble pigments occurring in cell sap. Widely distributed in leaves, flowers and stems. Some occur free and are yellow in colour but many are combined as glycosides (q.v.) and are practically colourless. Rarely colour flowers but modify other colours especially those of anthocyanins.

ANTHOZOA. Actinozoa (q.v.).

ANTHROPOID APES. The most nearly related to man of all living animals: gibbons, orang, chimpanzee, gorilla. They are Catarrhine Primates, forming the family Pongidae, all inhabiting Old World.

ANTHROPOIDEA (PITHECOIDEA). Sub-order of the order Primates. Includes monkeys, anthropoid apes, and man. Divided into two groups, Platyrrhini and Catarrhini. Large eyes, facing forwards: strong tendency to use hands for manipulation; relatively large brain.

ANTIBIOTIC. Substance produced by living organism which diffuses into its surroundings and is toxic there to individuals belonging to other species. E.g. penicillin, produced by *Penicillium notatum*, antagonizes many species of bacteria. Antibiotics may be important features of the competition amongst soil fungi and bacteria.

ANTIBODY. A protein produced in an animal when a certain kind of substance (an antigen) which is normally foreign to its tissues gains access to them; the antibody combines chemically with the antigen. Antibodies tend to be highly specific, in that they combine only with antigens of a particular kind. Antigens are mostly proteins or carbohydrates; and the specificity of their reaction is due to the structure of certain small areas on the surface of these molecules: These active areas evoke antibodies carrying matching structures than in turn combine only with the active areas. Importance of antibody formation is as defence mechanism, of vertebrates and some invertebrates (e.g. insects), against invasion by parasites, particularly by bacteria and viruses. Antibodies are not found in plants. When parasites or their poisonous products enter the tissues the animal produces antibodies, which circulate dissolved in the body fluids. This antibody production is stimulated by antigens forming part of, or produced by, the parasites. A given parasite is likely to bear several antigens, some peculiar to its species or strain, and therefore to evoke several sorts of antibodies equally specific. In some cases different pathogens have the same antigen, so that immunity to one confers immunity to another, e.g. vaccinia and smallpox. The combination of antibody with antigen kills or immobilizes parasites, or makes them more susceptible to phagocytes (q.v.) or makes their poison innocuous. Antibodies are formed by plasma cells (q.v.); and once produced they are found mainly in the blood, and may persist there long after disappearance of the antigen, conferring immunity to a new infection by the same sort of parasite. Immunity to disease from vaccination or inoculation is due to antibodies. Substances quite unconnected with parasites can act as antigens when injected, e.g. almost any foreign protein, and the specificity of the reaction provides an extraordinarily sensitive test for different proteins.

ANTICLINAL. (Bot.). (Of planes of division of cells) situated approximately at right angles to outer surface of plant part. Cf. *Periclinal.*

ANTICOAGULANT. Substance which prevents the clotting of blood, e.g. heparin. See *Saliva.*

ANTIDIURETIC HORMONE. Hormone secreted by posterior lobe of pituitary. A polypeptide. Stimulates water reabsorption by uriniferous tubule (q.v.), hence diminishes volume of urine. Deficiency causes diabetes insipidus.

ANTIGEN. Substance capable of stimulating formation of an antibody (q.v.).

ANURA (SALIENTIA). Frogs and toads. An order of the class Amphibia. Long hind legs for jumping; no tail; soft skin important in respiration, with no scales.

ANUS. The opening of the alimentary canal to the exterior through which undigested remains of food, bacteria associated with them, and solid excretions, are expelled. This is its only function, except in some aquatic animals which use hind-end of gut for respiration (e.g. some dragon-fly larvae). Present in most Metazoa but absent in all Platyhelminthes and Coelenterata. See *Cloaca.*

AORTA. In man and other mammals, the great artery which leaves the heart (from the left ventricle), and through which passes the arterial blood supply for the whole body (at about four litres per minute in man) to be distributed via the numerous arteries which branch off the aorta. See *Aorta, Dorsal.*

AORTA, DORSAL. Artery of vertebrates through which passes arterial blood to all the body, except the head and in some species the front limbs. Lies just below vertebral column, throughout most of the latter's length. In tetrapod vertebrates it is the continuation of the systemic arch (in mammals this arch, and sometimes the dorsal aorta too, is called the aorta). In fish it arises from the arteries drawing blood from the gills (efferent branchial arteries). In its course it gives off several large branches to viscera and limbs, and eventually in the tail becomes the caudal artery.

AORTA, VENTRAL. Large artery of fish and embryonic amniotes, which leads from ventricle of heart, and gives off branches to gills or aortic arches.

AORTIC ARCHES. Paired arteries (usually six) of vertebrate embryos, connecting ventral aorta with dorsal aorta by running up between gill slits or gill-pouches on each side, one in each visceral arch. In fish, develop into arteries carrying blood to and from gills. In tetrapods the two most anterior pairs (between mouth and the pouch equivalent to spiracle; and between latter and second pouch) disappear. Third pair forms proximal part of carotid arteries, fourth becomes systemic arch (on one side only, in birds and mammals), fifth disappears, and sixth forms part of pulmonary arteries.

APETALOUS. Lacking petals, e.g. wood anemone.

APHANIPTERA. Fleas. Order of endopterygote insects. Wingless; parasites of mammals and birds whose blood they suck; strong legs, good jumpers; larvae grub-like.

APHETOHYOIDEA. Placodermi (q.v.).

APHIS. Green flies, etc.; genus of hemipteran insects which suck plant juices. Some carry virus diseases and as such are of great economic importance.

APHYLLOUS. Leafless.

APICAL MERISTEM. Growing point at tip of root and stem in vascular plants, having its origin in a single cell, e.g. Pteridophyta, or in a group of cells, e.g. Spermatophyta. In the latter the apex of the growing point (*promeristem*) consists of actively dividing, uniformly shaped cells. Behind the promeristem differentiation begins, becoming progressively greater towards the mature tissues. One (older) concept of the organization of the growing point in flowering plants recognizes differentiation into three regions (*histogens*) behind the promeristem, *dermatogen*, a superficial layer of cells giving rise to the epidermis, *plerome*, a central core of tissue that gives rise to the vascular cylinder and pith, and *periblem*, tissue lying between dermatogen and plerome, that gives rise to cortex. It is now evident that the respective roles assigned to these histogens are by no means universal, nor is it always possible to distinguish between periblem and plerome, especially in the shoot apex. Now becoming widely accepted is the *Tunica-Corpus* concept, an interpretation of the shoot apex which recognizes two tissue zones *tunica*, consisting of one or more peripheral layers, in which the planes of cell division are predominantly anticlinal, enclosing *corpus*, or central tissue of irregularly arranged cells in which the planes of cell division vary. No relation is implied between cells of these two regions and differentiated tissue behind apex as in histogen concept. Although epidermis arises from outermost tunica layer, underlying tissue may originate in tunica or in corpus, or in both, in different plant species.

APLANOSPORE. Non-motile fungus spore formed within a sporangium. Characteristic of a group of Phycomycete fungi, the Zygomycetes (pin-moulds).

APOCARPOUS. (Of the gynoecium of flowering plants) having separate carpels, e.g. buttercup. See *Flower*.

APODA (GYMNOPHIONA). An order of Amphibia. Tropical, burrowing, limbless, wormlike. Unlike all other living Amphibia, have scales, buried in the skin.

APOGAMY. See *Apomixis*.

APOMICT. Plant produced by *Apomixis* (q.v.).

APOMIXIS. Reproduction which has superficial appearance of ordinary sexual cycle but actually occurs without fertilization and/or meiosis. Usually taken to include *parthenogenesis* (q.v.); *apospory*, in some plants, where an ordinary diploid sporophyte cell substitutes for the spore, meiosis being omitted, so that the gametophyte is diploid; and *apogamy*, in Pteridophytes, in which a gametophyte cell gives rise directly to a sporophyte, the gametophyte being diploid. Has also been applied recently to include vegetative reproduction whenever the normal sexual processes are not functioning or are greatly reduced in activity. See *Agamospermy*.

APOSEMATIC COLORATION. Warning Coloration (q.v.).

APOSPORY. See *Apomixis*.

APOTHECIUM. Cup or saucer-shaped fruit body of certain ascomycete Fungi (Discomycetes or cup-fungi), and Lichens, lined with a hymenium of asci and paraphyses. Sessile or stalked, often brightly coloured; varying from a few millimetres to more than forty centimetres across.

APPENDAGE. Any considerable projection from the body of an animal. *Paired a.* One of a bilaterally symmetrical pair of appendages. Two such pairs occur in all gnathostome vertebrates (but see *Acanthodii*). There is frequently one pair per segment in arthropods (e.g. walking-legs, mouth-parts, antennae) and annelids.

APPENDIX, VERMIFORM. Small diverticulum of caecum (at junction of large and small intestines) of man, apes, and some other mammals, containing much lymphoid tissue.

APPOSITION. (Bot.). Growth in thickness of cell walls by successive deposition of material, layer upon layer. Cf. *Intussusception*.

APTERYGOTA (AMETABOLA). Sub-class of wingless insects, including orders Thysanura (bristle-tails, silver-fish), Collembola (spring-tails), and Protura. It is believed that they are primitively wingless as distinct from other wingless insects, e.g. fleas, which are derived from winged ancestors. Cf. *Pterygota*.

AQUEOUS HUMOUR. Fluid which fills space between cornea and vitreous humour of the vertebrate eye. The iris and lens lie in it. Much like cerebrospinal fluid in composition. It is continuously secreted and absorbed. See Fig. 3, p. 86.

ARACHNIDA. Class of Arthropoda containing scorpions, spiders, ticks, mites, harvest-men, king-crabs. Most living members are terrestrial, breathing air, but king-crabs are aquatic. No antennae; first pair of appendages used for grasping; second may be grasping, sensory, or locomotory; remaining four pairs locomotory.

ARCHAEOPTERYX AND ARCHAEORNIS. Earliest known fossil birds (Jurassic, 140–170 million years ago), with teeth, claws on three-fingered hands, long tail containing numerous vertebrae; in many respects extremely like a reptile but with feathers (which are well preserved in the fossils).

ARCHEGONIATAE. Members of Cryptogamia (q.v.) in which female sex organ is an archegonium (q.v.), i.e. Bryophyta and Pteridophyta.

ARCHEGONIUM. Female sex organ of liverworts, mosses, ferns, and related plants, and most gymnosperms; multicellular, consisting of a neck, composed of one or more tiers of cells, and a swollen base (venter), containing the egg-cell.

ARCHENTERON. Cavity within early embryo (at gastrula stage) of many animals, communicating with exterior by blastopore (q.v.). Formed by invagination of mesoderm and endoderm cells during gastrulation. Ultimately becomes gut cavity.

ARCHESPORIUM. Cells or cell from which spores are ultimately derived, e.g. in developing pollen sac, fern sporangium.

ARCHIANNELIDA. Class of Annelida consisting of a few genera of small marine worms of simplified structure, possibly derived from Polychaeta.

ARCHICHLAMYDEAE. Sub-class of Angiospermae in which the individual members of the corolla are entirely separate from each other or the perianth is incomplete. Cf. *Sympetalae*.

ARIL. Brightly coloured, succulent investment of the seed in a few plants, e.g. yew, developed from stalk or base of ovule.

ARISTA. Awn (q.v.).

ARTEFACT. Something which does not occur in the undisturbed living tissue or organism, but was produced in it during investigation, or during its preparation for investigation (especially during fixation and dehydration). E.g. the meshwork seen in a fixed nucleus is an artefact of fixation not present in living nucleus.

ARTERIOLE. Small artery of vertebrates, less than $\frac{1}{3}$ mm. diameter. Smooth muscle of wall well developed, and under control of autonomic nervous system. Acts as a kind of stopcock regulating blood flow through capillaries (which are continuations of the arterioles).

ARTERY. Blood vessel carrying blood from the heart towards the tissues. In vertebrates arteries have thick walls, containing elastic and collagen fibres and smooth muscle, to withstand the high blood pressure near the heart. They are lined with smooth flat cells (endothelium), like other blood vessels. Arterial system starts with one or few arteries at heart, which by repeated branching as they proceed away from the heart give rise to an increasing number of even smaller arteries, until the branches become arterioles (q.v.).

ARTHROPODA. The largest phylum in the animal kingdom in number of species, including crabs, insects, spiders, centipedes, etc. Hard jointed exoskeleton; paired jointed legs; coelom reduced; no nephridia; no cilia. Contains classes Trilobita, Crustacea, Myriapoda, Arachnida, Insecta. See also *Onychophora.*

ARTICULATE WITH. Be attached by means of a movable joint to: e.g. the arm articulates with the shoulder girdle.

ARTIFICIAL INSEMINATION. Artificial injection of semen into female. Much used in animal breeding, since sperm of animal with desirable hereditary qualities can be transported over long distances and used to inseminate numerous females, under conditions optimum for fertility.

ARTIFICIAL PARTHENOGENESIS. Artificial activation of development of an egg (which is not normally parthenogenetic) without contact with sperm. This can be done in many animals (including rabbits) by treating eggs in various ways, e.g. cooling; pricking with a needle; treating with acid. In a small proportion of instances the activated egg has been reared to an adult. See *Parthenogenesis.*

ARTIODACTYLA. Even-toed ungulates. Order of mammals containing pigs, hippopotami, camels, and ruminants (deer, giraffes, sheep, goats, antelopes, oxen). Walk on their toes, on which there are hooves, and are characterized by fact that weight-bearing axis of the foot lies between third and fourth toe. The number of toes may be four (pigs, hippopotami) or more usually two, the typical 'cloven hoof'. One of the two great groups of hoofed mammals (ungulates), the other being the Perissodactyla. See also *Ruminant.*

ASCIDIAN. Sea-squirt; a member of the Urochordata.

ASCOCARP. General term for fruit body (cleistocarp, apothecium, or perithecium) of an ascomycete fungus.

ASCOMYCETES. A group of fungi (q.v.).

ASCORBIC ACID (VITAMIN C). Vitamin required by man and other primates and by guinea pigs, but synthesized by other animals studied. Water soluble. Deficiency causes scurvy, and prevents collagen-formation (e.g. in healing wounds). Widely present in organisms.

ASCOSPORE. See *Ascus.*

ASCUS. Characteristic spore-producing cell of ascomycete Fungi; almost spherical, club-shaped or cylindrical cell containing usually eight haploid spores (ascospores) formed following meiosis.

ASEXUAL REPRODUCTION. Reproduction without gametes (excludes parthenogenesis). In plants consists of either spore-formation or vegetative reproduction (q.v.). In Metazoa spore-formation does not occur; asexual reproduction is by fission (q.v.) or gemmation (q.v.).

ASSIMILATION. Absorption and building up of simple food-stuffs, or products of digestion of food-stuffs, into complex constituents of the organism.

ASSOCIATION. (Of plants), largest natural vegetation group that is recognized. Stable plant community dominated by particular species and named according to them, e.g. deciduous forest association, heath association. This, the original conception of the meaning of association, is not now universally held; e.g. Swedish and continental workers apply the term to very small natural units of vegetation.

ASSORTATIVE MATING (A. BREEDING). Sexual reproduction in which the pairing of male and female is not at random, but involves a tendency for males of a particular kind to breed with females of a particular kind: e.g. the two parents of each pair may tend to be more alike than is to be expected by chance.

ASTER. A system of striations in the cytoplasm, radiating from the centriole, often conspicuous during cleavage of the egg, or during fusion of nuclei at fertilization; also probably present in many other dividing animal cells. Absent in higher plants.

ASTEROIDEA. Starfishes. Class of Echinodermata. Star-shaped, arms not sharply marked off from central part of body.

ATAVISM. Recurrence in descendants of a character which had been possessed by an ancestor, after an interval of several or many generations.

ATLAS. First vertebra of tetrapods, modified for joint with skull. In, amniotes, which have freer head movement, it is modified further than in Amphibia; it consists of a simple bony ring, and there is a peg (odontoid process) on the next vertebra (axis) which projects forwards into the ring (through which spinal cord also runs). This peg represents part of the atlas (its centrum), which has become detached and fused to the axis. Nodding the head takes place at skull-atlas joint; rotating head at atlas-axis joint.

ATP (ADENOSINE TRIPHOSPHATE). A co-enzyme of many reactions. One of the phosphate groupings of ATP is readily transferred to other substances by enzyme action, simultaneously transferring a considerable amount of energy. This transfer of phosphate from ATP seems to be the main mechanism by which organisms make energy available for chemical synthesis, muscular contraction, osmotic work, etc. ATP is re-formed from adenosine diphosphate as a result of catabolic processes, the energy of which is thus made available for various kinds of work through ATP. See *Phosphagen.*

ATRIUM. Chamber, closed except for a small pore, surrounding gill-slits of Amphioxus and Urochordata. Also synonym of Auricle (q.v.).

ATROPHY. Diminution in size of an organ, or in amount of a tissue or constituent of a tissue.

AUDITORY (OTIC) CAPSULE. Part of skull of vertebrates enclosing auditory organ.

AUDITORY NERVE (ACOUSTIC NERVE). Eighth cranial nerve of vertebrates, innervating inner ear. A dorsal root, embryologically a sensory branch of seventh cranial nerve.

AUDITORY ORGAN. Sense-organ for detecting sound. In vertebrates the auditory organ (see *Ear, inner*) also detects position in relation to gravity and acceleration.

AURICLE (ATRIUM). One of the chambers of the heart; receives blood from veins and passes it to ventricle (q.v.). Has muscular walls, but is not as powerful a pumping organ as the ventricle. A fish has a single auricle, but in land-living (tetrapod) vertebrates, which breathe mainly or entirely by lungs, there are two auricles, one receiving oxygenated blood from lungs, other deoxygenated from the rest of the body. Most Mollusca have two auricles, one receiving blood from each side of the body. (Bot.) Small ear- or claw-like appendage occurring one on each side at the base of leaf-blades in certain plants.

AUSTRALOPITHECINE. Member of a group of fossil primates from South Africa. Distinctly human in some features, especially of limbs and teeth, but ape-like in others, especially of skull. Walked upright. Early Pleistocene.

AUTECOLOGY. Ecology of individual species as opposed to communities (*synecology*).

AUTOCATALYTIC. Catalysing its own production; so that the more that is produced, the more catalyst there is for further production.

AUTOECIOUS. Of rust Fungi (order Uredinales of Basidiomycetes), having the different spore forms of the life-cycle all produced on one host species, e.g. mint rust. Cf. *Heteroecious*.

AUTOGAMY. A curious sexual process found in some Protozoa in which the nucleus of an individual divides into two parts which reunite. Also occurs in some Diatoms. Sometimes used for self-fertilization in plants.

AUTOGRAFT. Graft originating from the individual which receives it. Cf. *Homograft, Heterograft*.

AUTOLYSIS. Self-dissolution that tissues undergo after death of their cells, due to action of their own enzymes.

AUTONOMIC. (Of plant movements) arising as a result of internal stimuli, e.g. nutation (q.v.). Alternatively described as autogenic or spontaneous. Cf. *Paratonic*.

AUTONOMIC NERVOUS SYSTEM. (1) Of vertebrates, motor nerve supply to smooth muscles (e.g. in gut, blood-vessels, etc.) and glands. A characteristic feature is that it does not involve single nerve fibres running all the way from central nervous system to effector (as with supply to striped muscles); but the nerve fibres (*preganglionic*) which leave C.N.S. stop short of effectors, and form synapses with a second group of nerve cells whose nerve-fibres (*postganglionic*), which are mainly unmyelinated, then continue to effectors. The cell-bodies of the postganglionic fibres, and the synapses, occur in ganglia or in the organ where the effectors are. The system is subdivided into *sympathetic* (or *orthosympathetic*) system, and *parasympathetic* system. In mammals and birds (other vertebrates are little known) preganglionic fibres of sympathetic system leave spinal cord (in thoracic and lumbar region) through ventral roots, and go to chains of *sympathetic ganglia*, one on each side just ventral to vertebral column, where they meet postganglionic nerve-cells, or pass through and meet them in another, median, row of ganglia (e.g. the solar plexus). From all these ganglia postganglionic fibres go to effectors. Preganglionic fibres of parasympathetic system on the other hand leave C.N.S. (*a*) through cranial nerves, especially vagus, (*b*) through ventral roots of hind end of spinal cord (sacral region), and the postganglionic cell-bodies are scattered amongst the effectors or are in ganglia very close to the effectors. The sympathetic system is the larger of the two; it alone supplies skin and limbs. Many internal organs receive nerve-fibres from both systems, and in such cases they may act antagonistically; e.g. smooth musculature of gut is stimulated to peristalsis by parasympathetic supply (mostly from vagus), inhibited by sympathetic supply. Parasympathetic fibres are cholinergic; sympathetic often adrenergic, sometimes (e.g. those to sweat glands) cholinergic. Much of the co-ordination of activities of the autonomic system occurs in spinal cord, medulla, and hypothalamus. (2) The motor system as described under (1), together with the sensory fibres from internal receptors which are concerned in reflex activities of this motor system.

AUTOPOLYPLOID. Polyploid (q.v.) in which all the chromosomes come from the same species. Cf. *Allopolyploid.*

AUTORADIOGRAPH. Photographic picture of the localization of radio-active substance in tissues, obtained by laying photographic emulsion in the dark on a thin preparation (usually a section) of the tissue and developing the image produced by the radioactivity. Commonly used in tracer work.

AUTOSOME. Chromosome which is not a sex-chromosome.

AUTOSTYLIC JAW-SUSPENSION. Attachment of primitive upper jaw (palatoquadrate) of vertebrates direct to neurocranium (see *Hyostylic Jaw-Suspension*). Occurs in the most primitive fish (Placodermi) which were probably succeeded by hyostylic descendants. Autostyly was then reacquired independently by Holocephali (q.v.) and by Dipnoi and tetrapods; in these, the palatoquadrate actually fuses with the brain-case.

AUTOTOMY. Self-amputation of part of the body, e.g. lizard can break off tail when it is seized by a predator, muscular action snapping a vertebra. Lost part is usually regenerated.

AUTOTROPHIC. (Of an organism) independent of outside sources of organic substances for provision of its own organic constituents, which it can manufacture from inorganic material. An autotrophic organism may be phototrophic (q.v.) or chemotrophic (q.v.) as regards its supply of energy. Most chlorophyll-containing plants are autotrophic, manufacturing organic materials from water, carbon-dioxide, nitrates, and other salts with sunlight as the phototrophic source of energy (see *Chlorophyll*). A few bacteria are autotrophic, using inorganic materials, and chemotrophic, using energy produced by inorganic oxidations, e.g. of hydrogen sulphide by sulphur bacteria or of hydrogen by hydrogen bacteria. All other organisms, which are *heterotrophic*, depend ultimately on the synthetic activities of autotrophic organisms. See *Metabolite*.

AUXIN. See *Hormone*.

AUXOTROPH. Strain of a micro-organism (alga, bacterium or fungus) which requires growth factors not needed by wild type (q.v.) or by prototrophs (q.v.). Cf. *Prototroph*.

AVES. The birds, a class of vertebrates. Characterized by feathers, warm blood, wings developed from fore-limbs. Apart from some fossil forms (e.g. *Archaeopteryx*) birds form a highly uniform class. Descended from reptiles related to dinosaurs and crocodiles; and still have many reptilian features.

AWN (ARISTA). Stiff, bristle-like appendage occurring frequently on the flowering glumes of grasses and cereals.

AXIL. (Of a leaf) angle between its upper side and the stem on which it is borne; normal position for lateral buds.

AXIS CYLINDER. Axon (q.v.).

AXOLOTL. Aquatic larval stage of an American salamander (genus *Ambystoma*, Order Urodela) which is remarkable in that it does not metamorphose, but breeds while keeping its larval form (the phenomenon of neoteny, q.v.). Such breeding axolotls occur only in some districts; in others, *Ambystoma* metamorphoses and breeds as normal adult form.

AXON. The long process of a nerve cell, normally conducting impulses *away from* the nerve cell-body (cf. *Dendrite*). Often covered by myelin. See *Nerve-Fibre*.

B

BACILLARIOPHYCEAE. Diatoms. Class of Algae. Microscopic, unicellular plants, occurring singly or grouped into colonies. Chloroplasts contain brown pigment isofucoxanthin in addition to chlorophyll. Abundant in marine and freshwater plankton. Cell wall of pectic materials impregnated with silica, finely sculptured; composed of two halves, one of which overlaps the other. Asexual reproduction by cell division, sexual reproduction isogamous. Past deposition of countless numbers of the silicified cell walls has formed the deposits known as siliceous or diatomaceous earths.

BACILLUS. Genus of spore-producing bacteria. Also used generally for rod-shaped bacteria.

BACKBONE. Vertebral column (q.v.).

BACKCROSS. Cross of a hybrid with one of its parents.

BACTERIA. Ubiquitous, unicellular or multicellular, microscopic organisms with simple nucleus, lacking chlorophyll. There is no general agreement as to the systematic position of bacteria but they are usually included in the division Thallophyta (q.v.) of the plant kingdom. In shape rod-like, more or less spiral, filamentous, occasionally forming a mycelium, or, in some forms, spherical, varying in breadth mostly from 0.5 to 2.0 microns; non-motile or motile by one or more flagella. Multiplication is by simple fission; other forms of asexual reproduction, by formation of aerially dispersed spores, flagellated swarmers, occur in some bacteria. Sexual reproduction has also been demonstrated in certain forms. Occurring in large numbers in favourable habitats, e.g. a gram of soil may contain from a few thousand to several hundred million bacteria; a cubic centimetre of sour milk, many millions. Most bacteria are saprophytes or parasites. A few are autotrophic (q.v.), either obtaining energy by oxidation processes or from light in the presence of bacteriochlorophyll. The activities of bacteria are of great importance. In the soil they are concerned in the decay of plant and animal tissues, making available food materials for higher plants, and certain bacteria are of particular significance in this respect in making available nitrogenous compounds. See *Nitrogen Cycle*. Others are sources of antibiotics, e.g. *Streptomyces griseus*, source of streptomycin. Bacteria also play an important part in the disposal of sewage. As agents of plant disease, bacteria are not so important as fungi, but they cause many serious diseases in animals and man, e.g. tuberculosis, diphtheria, typhoid, pneumonia.

BACTERIOCHLOROPHYLL. See *Chlorophyll*.

BACTERICIDAL. Lethal to bacteria. Cf. *Bacteriostatic*.

BACTERIOLOGY. Study of bacteria.

BACTERIOPHAGE ('PHAGE). A virus that destroys bacteria.

BACTERIOSTATIC. Inhibiting growth of bacteria but not killing them. Cf. *Bactericidal*.

BALANOGLOSSUS. Genus of worm-like burrowing animals belonging to the Hemichordata (q.v.), with vertebrate affinities.

BALEEN. Whalebone. Transverse plates of keratin, derived from epidermis, hanging from the upper jaw on each side of the mouth of the toothless whalebone whales (rorquals and right whales). Frayed inner edges form a filter, retaining the small animals on which the whale feeds.

BARBS. Of feathers. The filaments, in a row at each side of the longitudinal axis, which together make the expanded part (vane) of a feather. See *Barbules.*

BARBULES. Of feathers. Minute filaments in a row at each side of a barb. Those of one side bear hooks, those of the other a groove. Barbules of adjacent barbs hook together and link barbs into a firm vane. Down-feathers and ostrich feathers have no interlocking mechanism, so their barbs are free.

BARK. Protective, corky tissue of dead cells, present on the outside of older stems and roots of woody plants, e.g. tree trunks, produced by activity of cork cambium. Bark may consist of cork only, or, when other layers of cork are formed at successively deeper levels, it may consist of alternating layers of cork and dead cortex or phloem tissue (when it is known as *rhytidome*). Popularly regarded as everything outside the wood

BASEMENT MEMBRANE. Delicate intercellular membrane which underlies most animal epithelia. Usually contains polysaccharide and very fine (reticulin) fibres.

BASIC DYES. Dyes consisting of a basic organic grouping of atoms (cation) which is the actively staining part, combined with an acid, usually inorganic. Stain particularly nucleic acids, and therefore nuclei. Cf. *Acid dye.*

BASIDIOMYCETES. A group of fungi (q.v.).

BASIDIOSPORE. See *Basidium.*

BASIDIUM. Characteristic spore-producing structure of basidiomycete Fungi. Single, club-shaped or cylindrical cell, or divided into four cells, and bearing externally on minute stalks (sterigmata) four spores (basidiospores).

BASIPETAL. (Bot.). (Of organs), development in succession towards the base, oldest at the apex, youngest at base. Cf. *Acropetal.*

BASOPHILIC. Staining strongly with basic dye. Especially characteristic of nucleic acids, and hence of nucleus (during mitosis of which, basophily is localized in chromosomes) and of cytoplasm when actively synthesizing proteins.

BATRACHIA. Amphibia (q.v.).

BENNETTITALES. Order of extinct Gymnospermae (q.v.) that flourished during the Mesozoic (q.v.). Resembled cycads (Cycadales) somewhat in external appearance and anatomy of stem and leaf. Differed from them in possessing cones containing both micro- and megasporophylls and in form of these sporophylls.

BENTHOS. Those animals and plants living on the bottom of sea or lake, from high water mark down to the deepest levels. The benthos is divided into littoral (down to 200 metres deep) and deep water organisms. See *Pelagic, Nekton, Plankton.*

BERRY. Many-seeded succulent fruit in which wall (*pericarp*) consists of outer skin (*epicarp*), comparatively thick fleshy mesocarp and inner membranous *endocarp*, e.g. gooseberry, currant, tomato. Cf. *Drupe*.

BICOLLATERAL BUNDLE. See *Vascular Bundle*.

BIENNIAL. Plant that requires two years to complete its life-cycle, from seed germination to seed production and death. During the first season biennials, e.g. carrot, cabbage, store up food which is used in the second season when they produce flowers and seed. Cf. *Annual, Ephemeral, Perennial*.

BILATERAL CLEAVAGE. Cleavage (q.v.) which produces a bilaterally symmetrical arrangement of blastomeres lacking the peculiar arrangement and oblique divisions of spiral cleavage (q.v.). Occurs in Echinodermata, Chordata, etc.

BILATERALLY SYMMETRICAL. Capable of being halved in one, and only one, plane in such a way that the two halves are approximately mirror-images of each other. Usually this plane lies antero-posteriorly and dorso-ventrally, thus separating similar right and left halves. Almost all freely moving animals are bilaterally symmetrical, e.g. all vertebrates, arthropods, and worm-shaped animals. Similar condition in flowers, e.g. snapdragon, is often called *zygomorphy*. See *Radially Symmetrical*.

BILE. Secretion of liver of vertebrates, passed through bile-duct to duodenum. Important in digestion of fats, which, through action of bile-salts, are broken into minute droplets (emulsified). Contains also pigments which are waste-products of haemoglobin destruction.

BILE DUCT. Duct from liver to duodenum of vertebrates. Conveys bile. See *Gall-Bladder*.

BILE SALTS. Sodium salts of taurocholic and glycocholic acids (bile acids, see *Steroids*), secreted in bile. Strongly lower surface tension, emulsifying fat. Responsible for bitter taste of bile.

BILHARZIA. Schistosoma (q.v.).

BINOCULAR VISION. Type of vision occurring in primates and many other vertebrates, especially active predators, in which eyeballs can be so directed that image of an object falls on both retinas. Extent to which eyes must be turned to bring the images on to a special part of the retinas (see *Fovea*) is probably one of the mechanisms of distance judgement. Stereoscopic vision (perception of shape in depth) depends on two slightly different images being received, because the two eyes look from different angles.

BINOMIAL NOMENCLATURE. The present method of naming species of animals and plants scientifically. When a new organism is discovered a description of it is published, often in Latin, and it is given a name, always in Latin. The name is in two parts (binomial): one part (the specific epithet or trivial name) peculiar to the new species; the other part (the generic name) designating the genus (q.v.) to which it belongs, and therefore applied in the same way to other closely related species, if there are any, but to no other genus. The generic name is written first with a capital letter, the specific second, usually with a small letter, e.g. *Felis leo* (the lion) is one species, *Felis tigris* (the tiger) is another of the same genus. Scientific names are usually written in italics. Strictly the name (sometimes abbreviated) of the author responsible for naming and describing the species should follow, e.g. *Felis leo* Linn. (the author here being Linnaeus). When subsequent to the original description the species has been transferred to a different genus, the original author's name is put into brackets. It is usual for one specimen to be designated as the 'type' of a given species; in any later splitting of the species the original specific epithet must continue with the type and those specimens like it (see *Type Specimen*). These and other rules form part of the International Rules of Nomenclature, with separate rules for Botany and Zoology. Naming is supervised by an international committee. The binomial system was introduced by the early herbalists, and was first systematically applied by Linnaeus in the middle of the eighteenth century. The names used by Linnaeus in the *Species Plantarum* (1753) and in the tenth edition of his *Systema Naturae* (1758) are the basis of the system for plants and animals respectively, apart from a few groups with which he did not deal, for which other classical descriptive works serve. Species not included in these are given the epithet first applied to them. If such an epithet is later found not to have been the first given to the species, it must, according to the rules of priority, be replaced by the earlier one, which leads to some instability of nomenclature. Provision is however made in special cases for waiving the rules. Where sub-species or varieties have been defined, a trinomial system is commonly used, the third name indicating the sub-species or variety, e.g. *Troglodytes troglodytes troglodytes*, the British wren.

BIO-ASSAY. Quantitative estimation of biologically active substances by the amount of their actions in standardized conditions on living organisms, e.g. androgen estimation on capon's comb.

BIOCHEMISTRY. Study of chemical substances and chemical processes of living things.

BIOGENESIS, PRINCIPLE OF. The biological rule that a living thing can originate only from a parent or parents on the whole similar to itself. It denies both spontaneous generation (q.v.) and such fanciful notions current 300 years ago as the origin of geese from barnacles. If allowance is made for alternation of generations, complex life cycles, extreme sexual dimorphism, castes, and hybridization, it is true; though it may possibly need further qualification in the case of the simpler viruses, which it has been suggested may arise from normal constituents of cells.

BIOLOGICAL CONTROL. Control of pests and parasites by use of other organisms, e.g. of mosquitoes by fishes or by insectivorous plants which feed on the larvae; or of prickly pear in Australia by parasitic insects.

BIOLOGIC SPECIALIZATION. See *Physiologic Specialization.*

BIOLOGY. Study of living things.

BIOLUMINESCENCE. Production of light by living organisms, e.g. fireflies, numerous marine animals, many bacteria and fungi. The light is due to an enzyme-catalysed chemical reaction which produces very little heat. It is of various colours. In animals may be under nervous control. See *Phosphorescence.*

BIOMETRY. Application of mathematics to the study of living things; particularly statistical study of resemblances and differences between groups of related organisms.

BIONOMICS. (Zool.). Study of the relation of an organism or population of organisms to its environment, animate and inanimate.

BIOPHYSICS. Application of physics to the study of living things.

BIOSPHERE. That part of the earth and its atmosphere which consists of living things.

BIOTIC FACTORS. Environmental influences that arise from activities of living organisms, as distinct from e.g. climatic factors.

BIOTYPE. (1) Naturally occurring group of individuals having the same genetic composition. (2) Physiologic race, see *Physiologic specialization.*

BIRAMOUS APPENDAGE. Forked appendage of Crustacea. The two branches may be similar, e.g. abdominal appendages of crayfish, or dissimilar.

BISEXUAL. Hermaphrodite (q.v.).

BIURET REACTION. Biochemical test used for proteins in solution. Employs copper sulphate in alkaline solution, which gives purple colour with proteins and with a few other substances.

BIVALENT. Two homologous chromosomes while they are pairing during meiosis (q.v.).

BIVALVE. Having a shell in two parts hinged together, e.g. lamellibranch Mollusca (mussels, etc.), Brachiopoda (lamp shells).

BLADDER. See *Gall Bladder, Swim Bladder, Urinary Bladder.*

BLADDERWORM. Cysticercus (q.v.).

BLASTEMA. (Zool.). Mass of undifferentiated cells which later develops into an organ. One of the two main ways by which an animal regenerates a lost part is by initial formation of a blastema (e.g. limb or tail of newt, head of flatworm); the other way being by remodelling of remaining tissue (morphallaxis).

BLASTOCOELE. Cavity which appears within the mass of cells formed towards end of period of cleavage of the egg of many animals. See *Blastula.*

BLASTOCYST. Stage in mammalian development resulting from cleavage, which is roughly the equivalent of a blastula. A thin-walled hollow sphere (trophoblast) with at one side a knob or sheet of cells destined to become the embryo proper.

BLASTODERM. Superficial sheet of cells formed as result of cleavage (q.v.) of a yolky egg. In a yolky vertebrate egg (such as hen's egg) the blastoderm is a flat disc of cells at one pole of the yolk. In an insect's egg it is a layer of cells which completely surrounds the internal mass of yolk.

BLASTOMERE. (Zool.). One of the cells formed from the fertilized egg during cleavage.

BLASTOPORE. (Zool.). Transitory opening on surface of embryo in gastrula stage, by which the internal cavity (archenteron) communicates with the exterior; produced by invagination (q.v.), of superficial cells in the course of movements of gastrulation (q.v.), to form endoderm and mesoderm. In many animals the blastopore becomes the anus; in others it closes up at the end of gastrulation, and the anus later breaks through at the same place or nearby. In some animals gastrulation movements carry endoderm and mesoderm cells inwards at a particular place without forming an open pore (e.g. in a primitive streak, q.v.); the place of inward migration may then be called a virtual blastopore. See *Organizer*.

BLASTULA. Stage of embryonic development of animals, at or near the end of period of cleavage, and immediately preceding gastrulation movements. Usually (in those animals with complete cleavage) consists of a hollow ball of cells.

BLENDING INHERITANCE. Inheritance, after crossing two parents widely divergent in some quantitative character, e.g. weight, such that most members of the F_1, F_2 (q.v.) and subsequent generations are approximately intermediate in character between the parents. Occurs when variation of the character is affected by a number of different genes, each with small and additive effect. Segregation of such genes results in preponderance of intermediate types, and rarity of extreme types resembling the parents. Does not involve blending of genes, as originally thought (proved by fact that F_2 is more variable than F_1; with blending of genes variability would diminish with each generation).

BLIND SPOT. Place where optic nerve enters retina of vertebrate eye; devoid of rods and cones and hence blind. See Fig. 3, p. 86.

BLOOD. A fluid of animals contained in vessels or spaces with endothelial walls, circulated by muscular action of vessels or specialized parts of them (hearts), usually containing respiratory pigment, and carrying oxygen, food-materials, excretions, etc., through the body. Present in Nemertea, Annelida, Arthropoda, Mollusca, Brachiopoda, Phoronidea, Chordata.

BLOOD CLOTTING (B. COAGULATION). Conversion of liquid blood to jelly, occurring when blood-vessels are injured. Prevents escape of blood. In vertebrates, a soluble blood-protein (fibrinogen) is converted to tangle of insoluble threads (fibrin) by an enzyme (thrombin). Thrombin is formed from a blood-protein (prothrombin) by an activator (thrombokinase) liberated from injured tissues or from blood platelets (q.v.); removal of free calcium from blood (as by adding oxalate or citrate) inhibits thrombin formation. In some Arthropoda clotting is by projection of thread-like processes from blood-corpuscles.

BLOOD CORPUSCLE. Cell which circulates in blood. Particularly abundant in vertebrates. See *Red Blood Cell, White Blood Cell.*

BLOOD FILM. Thin smear of blood, made on a slide, dried, fixed, and stained. Used for examination of corpuscles and detection of blood parasites.

BLOOD-GROUP. Group of people whose blood may be mixed without clumping of blood corpuscles. There are four main groups, A, B, AB, and O. Agglutination occurs when blood from any two different groups is mixed, owing to a reaction between substances (agglutinogens) in corpuscles, and other substances (agglutinins) in plasma. Group A has agglutinogen A in its corpuscles and agglutin anti-B (which reacts with agglutinogen B) in its plasma. Group B has agglutinogen B and agglutinin anti-A (which reacts with agglutinogen A). Group AB has both agglutinogens but neither agglutinin. Group O has neither agglutinogen but both agglutinins. Possession of agglutinins is not due to acquired immunity; every individual has the agglutinins which react with the agglutinogens he has not got. The four groups are due to different combinations of three multiple allelomorphs, g, I^A, I^B; g is recessive to the other two, which show no dominance to each other. Homozygous g is group O, homozygous I^A or heterozygous $I^A g$ are group A; homozygous I^B or heterozygous $I^B g$ are group B; heterozygous $I^A I^B$ is AB. Proportions of the four groups within population are very different in different parts of the world. From point of view of blood transfusion, introduction of corpuscles containing agglutinogen for which recipient has agglutinin must be avoided. The converse is not so important. Members of the population differ in other antigens carried in corpuscles (e.g. M and N antigens, forming M, MN, and N types); but since agglutinins for these are not normally present in plasma, they have not the same significance in transfusion, though immunity to them may be acquired by repeated transfusion. See *Rh factor.* Similar blood groups occur in other Primates.

BLOOD PLASMA. Blood from which all blood corpuscles have been removed, e.g. by centrifuging. A clear almost colourless fluid in vertebrates which clots as easily as whole blood. See *Plasma Proteins.*

BLOOD PLATELETS. Minute bodies, probably fragments of cells, in mammalian blood. Roughly 250,000 per cu. mm. of human blood. For function, see *Blood Clotting.* In other vertebrates represented by small spindle-shaped nucleated cells, *thrombocytes.*

BLOOD PRESSURE. Usually refers to pressure of blood in main

arteries, which in 'normal' human beings fluctuates roughly between 120 and 80 mm. of mercury according to stage of heart beat (maximum at systole, minimum at diastole). Pressures throughout the whole circulation diminish from arteries next to heart round to veins next to heart; in mammals reaching atmospheric or less in large veins.

BLOOD SERUM. Fluid expressed from clotted blood or from clotted blood plasma. Roughly, plasma deprived of clotting constituents.

BLOOD-SUGAR. Glucose circulating in blood. In vertebrates supplied to blood by liver (where it is stored as glycogen) and removed therefrom by all body cells for use as food. In mammals maintained at a fairly constant level (80–180 mg. per 100 cc. blood in man), chiefly controlled by hormones (insulin, adrenalin, and hormone from anterior pituitary). If level is too low (hypoglycaemia) the cells are starved, and this rapidly damages brain-cells; if too high (hyperglycaemia as in diabetes mellitus) glucose is excreted by kidney.

BLOOD-VESSEL. Tube through which blood flows. See *Artery, Arteriole, Capillary, Venule, Vein, Heart.*

BLUBBER. Thick layer of fatty tissue below dermis of skin, insulating against heat-loss, in aquatic mammals (whales, dugongs). See *Cetacea, Sirenia.*

BODY CAVITY. Internal cavity of most triploblastic animals (absent in Platyhelminthes, Nemertea) in which many organs are suspended, allowing their mutual displacement, which is especially important for the gut. Varies much in size in different groups. Bounded externally by *body wall.* Contains fluid. May develop embryologically in several ways, e.g. from coelom (earthworm, vertebrate), from blood system(Arthropoda) or as intercellular spaces (Nematoda). *Primary body cavity,* the haemocoel (q.v.). *Secondary body cavity,* the coelom (q.v.).

BONE. Skeletal substance peculiar to vertebrates. Consists of cells distributed in a matrix consisting largely of collagen fibres together with a complex salt (bone salt) mainly of calcium and phosphate. Bone salt, 60 per cent by weight of bone, is responsible for hardness; collagen for tensile strength. The cells are connected by fine channels which permeate the matrix. Larger channels contain blood-vessels and nerves.

BONY FISH. Osteichthyes (q.v.).

BOTANY. Study of plants.

BOWMAN'S CAPSULE. Part of Malpighian corpuscle of vertebrate kidney. Small sac ($\frac{1}{5}$ to $\frac{1}{10}$ mm. diameter in man); inward projection of its wall contains tuft of capillaries (glomerulus, q.v.); from it leads uriniferous tubule (q.v.).

BRACHIAL. (Adj.). Of the arm. *B. plexus.* Interconnexions formed in shoulder region between the spinal nerves supplying the fore-limb (arm) of tetrapods (fifth to ninth spinal nerves in man).

BRACHIOPODA (LAMP SHELLS). Phylum of animals with a two-valved shell, superficially resembling, but not closely related to, mussels, etc. During adult life fixed by a stalk or by one shell; feed by currents made by ciliated epithelium. Group is now small in numbers of individuals and of species, but was larger in Palaeozoic and Mesozoic times.

BRACT. Small leaf with relatively undeveloped blade, in axil of which arises a flower or a branch of an inflorescence.

BRACTEOLE. Small bract.

BRAIN. Anterior part of central nervous system, present in almost all bilaterally symmetrical animals, enlarged in connexion with aggregation of sense organs in head region. To a varying degree co-ordinates reactions of whole body.

BRAIN STEM. Vertebrate brain excluding cerebral hemispheres and cerebellum.

BRANCHIAL. (Adj.). Of the gills.

BRANCHIAL ARCH. Visceral arch lying between adjacent gill-slits of fish, i.e. the third and following visceral arches in most fish.

BRANCHIOPODA. Sub-class of Crustacea; aquatic; appendages uniform and usually flattened; abdomen without appendages; includes fairy shrimps, brine shrimps, water-fleas.

BRONCHIOLE. Small (less than 1 mm. diameter) air-conducting tube of tetrapod lung, arising as branch of a bronchus, terminating in alveoli. Smooth muscle abundant in walls, controlling size of lumen; the cartilage and mucus glands found in the bronchi are absent.

BRONCHUS. Large air-tube of tetrapod lung. Each lung has one large bronchus, connecting it to trachea; within the lung the bronchus branches into smaller and smaller bronchi, and finally into bronchioles. Have cartilage plates, smooth muscle, and mucus-secreting gland-cells in wall; and lining-cells bear cilia beating towards mouth, which remove dust, etc.

BRYOPHYTA. Division of plant kingdom comprising Hepaticae (liver-worts) and Musci (mosses). A small group of plants with wide distribution. Habitats various, e.g. wet banks, on soil, rock surfaces; some are epiphytes (q.v.), others aquatic. Small plants, flat, prostrate, or with a central stem up to a foot in length bearing leaves. Lacking vascular tissue. Attached to substratum by rhizoids. Reproducing sexually by fusion of male and female gametes produced in multicellular sex organs: antheridia, liberating male gametes motile by flagella, and archegonia, containing a single egg. Sexual reproduction is followed by development of a capsule, containing spores, which give rise, directly or indirectly, to new liverwort or moss plants. Bryophyta show a well-marked alternation of generations (q.v.). The plant itself is the gametophyte generation, and the capsule is the sporophyte generation, nutritionally dependent on the gametophyte.

BRYOZOA. Polyzoa (q.v.).

BUCCAL. (Adj.). Of the mouth. *B. cavity.* Mouth cavity lined by derivative of ectoderm. *B. membrane.* Embryonic membrane closing off the mouth, formed by ectoderm of buccal cavity externally, endoderm of pharynx internally.

BUD. Compact, undeveloped shoot, consisting of a short stem bearing crowded, overlapping, immature leaves.

BUDDING. (1) (Bot.). Form of grafting in which grafted part is a bud; (2) (Bot.). Asexual reproduction in which a new cell is formed as an outgrowth of a parent cell, e.g. yeast. Cf. *Fission*; (3) (Zool.). Gemmation (q.v.).

BUFFERED. Resisting changes in pH (q.v.) when acid or alkali is added. A property of many biological fluids, and of seawater.

BULB. Organ of vegetative reproduction; modified shoot consisting of a very much shortened stem enclosed by fleshy, scale-like leaves, e.g. tulip, or leaf-bases, e.g. onion. Cf. *Corm.*

BULBIL. Small bulb.

BULLA. See *Ear, Middle.*

C

CADUCOUS. (Bot.). Not persistent. Of sepals, falling off as flower opens, e.g. poppy; of stipules, falling off as leaves unfold, e.g. lime.

CAECUM. A blindly-ending branch of gut or other hollow organ. In amniote vertebrates and some fish there may be one or two caeca at the junction of small and large intestines; very large and important in digestion in some mammals, e.g. rabbit; vestigial in man.

CALCICOLE. (Of plants) growing best on calcareous soils.

CALCIFUGE. (Of plants) growing best on acid soils.

CALLOSE. Carbohydrate deposited on the sieve-plates of sieve-tubes (q.v.), bringing their activity to an end, either permanently or seasonally.

CALLUS. (Bot.). Superficial tissue developing in woody plants in response to wounding, usually by activity of cambium, protecting the injured surface. (Zool.). Material, at first containing collagen and cartilage, and later becoming bone tissue, which makes initial union of a broken bone.

CALORIE (=KILOCALORIE). (Written with capital C.) Amount of heat required to raise 1 kg of water from 15° to 16° C. One thousand times larger than the ordinary calorie of physics. Used as measure of energy turn-over in animals. E.g. daily energy output (as heat, work, and energy content of excretions) of an active man is on the average equivalent to about 3000 Calories, and food yielding that amount of heat when burned must be provided to maintain this output.

CALYPTRA. Hood-like covering of the capsule of mosses and liverworts, developed from archegonial wall.

CALYPTROGEN. Layer of actively dividing cells formed over apex of growing part of roots in many plants, e.g. grasses; gives rise to root-cap (q.v.).

CALYX. Outermost part of a flower, consisting usually of green, leaf-like members known as *sepals* that in the bud stage enclose and protect the other flower parts. See *Flower*.

CAMBIUM. A meristem (q.v.). Layer of actively dividing cells lying between xylem and phloem; forms additional xylem and phloem elements in the process known as *secondary thickening* (q.v.). The cambium of a vascular bundle is *fascicular* cambium; that formed from cells of parenchyma between vascular bundles and linking up with fascicular cambium to form a complete ring is *interfascicular* cambium.

CAMBRIAN. Geological period (q.v.); lasted approximately from 600 till 500 million years ago.

CAMPYLOTROPOUS. (Of ovule), curved over so that funicle appears to be attached to the side, between chalaza and micropyle. Cf. *Anatropous. Orthotropous.*

CANADA BALSAM. Gum commonly used, dissolved in xylene, for making permanent microscopical preparations. The object is placed in a thin layer of balsam solution between cover-slip and slide. The balsam dries hard, and because its refractive index is like that of proteins and other constituents of biological objects, it makes them very transparent.

CANINE TOOTH. 'Dog' or 'eye' tooth of mammals. Usually conical and pointed, one on each side of upper and lower jaws, between incisors and premolars. Missing or reduced in many rodents and ungulates. Sometimes enlarged to tusks (wild boar, sabre-toothed tiger).

CAPILLARY, BLOOD. Minute tube (very roughly 5–20 microns internal diameter; a given capillary can dilate or constrict) conveying blood, which it receives from a small artery (arteriole in vertebrates) and gives up to a small vein; with a wall consisting of a single layer of flattened cells (endothelium), supported by some fine connective tissue fibres. Capillaries in very large numbers permeate almost all the tissues. The main exchange of substances between blood and tissues occurs through capillary walls, which are permeable to small molecules (such as oxygen, glucose, amino-acids, which pass from blood to tissues; carbon dioxide which passes from tissues to blood). Walls are also freely permeable to water and salts. The reason why the water of the blood does not leak away through capillaries is that the large molecules, especially plasma proteins, are retained by capillary walls and hold back water by osmosis. In Arthropoda and Mollusca capillaries are physiologically largely replaced by haemocoel.

CAPILLITIUM. (1) Tubular protoplasmic threads in fruit bodies of Myxomycetes (Slime Moulds), shown to assist discharge of spores in some species by their movements in response to changes in humidity. (2) Sterile hyphae in fruit bodies of certain Fungi, e.g. puff-balls.

CAPITULUM. Kind of inflorescence (q.v.).

CAPSULE. (1) In flowering plants, dry, dehiscent fruit developed from a compound ovary (q.v.); opening to liberate seeds in various ways, e.g. by longitudinal splitting from apex to base, the separated parts being known as *valves*, as in iris; by formation of pores near top of fruit, as in snapdragon; or in the *pyxidium*, by detachment of a lid following equatorial dehiscence as in scarlet pimpernel. (2) In liverworts and mosses, organ within which spores are formed. (3) In some kinds of bacteria, gelatinous envelope surrounding the cell membrane. (4) (Zool.) Connective tissue investment of an organ, providing mechanical support.

CARAPACE. (1) Shield of exoskeleton covering part of the body (several segments) of some Arthropoda, e.g. crabs. (2) Dorsal part of 'shell' of Chelonia, consisting of exoskeletal plates fused with ribs and vertebral column.

CARBOHYDRATE. Compound of general formula $C_x(H_2O)_y$; e.g. sugars, starch, cellulose. Carbohydrates play an essential part in metabolism of all organisms. They are not present in nearly such large amounts in animals as in plants, in which cellulose is a principle structural component and starch the principal stored food.

CARBON CYCLE. World-wide circulation of carbon atoms brought about mainly by living things. Essentially, carbon from carbon dioxide is built into complex organic compounds by plants during photosynthesis. These compounds are then broken down again to carbon dioxide during respiration and by decay produced by bacteria and fungi after death; or the plants are eaten by herbivorous animals which in turn may be eaten by carnivores (see *Food-chain*) and the carbon compounds are sooner or later broken down to carbon dioxide by respiration or death and decay of the animals.

CARBONIFEROUS. Geological period (q.v.); lasted approximately from 350 till 270 million years ago.

CARCINOGEN. Producer of cancer, e.g. certain hydrocarbons which produce local cancer when injected into, or painted repeatedly on, susceptible animals.

CARDIAC. Concerning the heart. *C. cycle*, succession of muscular contractions, and movements of heart valves, which make up activity of heart from one beat to next. *C. end of stomach*, region of stomach joined by oesophagus. *C. muscle*, special kind of muscle, arranged in a meshwork of muscle fibres with cross-striations, which gives motive power to heart. Undergoes automatic rhythmical contractions.

CARDINAL VEINS. Main longitudinal veins, one on each side, returning blood from most of the body to ducti Cuvieri (q.v.) and thence to heart, in fish and embryos of tetrapods. *Anterior cardinals* (jugulars) are in front of ducti Cuvieri, *posterior cardinals* are behind. See *Vena cava inferior*.

CARINA. Keel (q.v.).

CARINATES. Living birds other than ratites. Most of them are capable of flight, with well-developed wings, barbules on feathers, and keel (carina) on sternum for attachment of wing muscles. The term is not now commonly used in classification of birds.

CARNASSIAL TEETH. Molar or premolar teeth modified for shearing flesh. In many Carnivora (e.g. dog, cat), first lower molar, and last upper premolar.

CARNIVORA. Order of placental mammals containing flesh-eating forms (cats, wolves, bears, seals). Most (but not seals) have well-developed incisor and canine teeth and usually a pair of carnassial teeth on each side. Claws, often retractile.

CARNIVORE. Flesh-eater. Cf. *Herbivore*, *Omnivore*. In narrower sense, member of order Carnivora.

CAROTENE. Photosynthetic orange pigment (hydrocarbon $C_{40}H_{56}$) occurring in chloroplasts, and in plastids in other plant parts where chlorophyll is absent, e.g. carrot roots, many flowers. Carotene of food is changed into vitamin A in vertebrate liver.

CAROTID ARTERY. Main artery supplying blood from heart to head of vertebrates. One on each side. See *Aortic Arches*.

CARPAL BONES. Bones of the proximal part of the foot of the fore-limb (roughly of the wrist) in tetrapod vertebrates. Compact group of primitively 10–12 bones; eight in man. Articulate on proximal side with radius and ulna, on distal side with metacarpals. See Fig. 7, p. 175.

CARPEL. Female reproductive organ of flowering plants, a megasporophyll; consisting of ovary containing ovules (q.v.) which become seeds after fertilization, and stigma (q.v.) a receptive surface for pollen grains, often borne at the apex of a stalk, the style. See *Flower*.

CARPOGONIUM. Female sex organ of red Algae. Consists of swollen basal portion containing the egg and an elongated terminal projection (trichogyne) which receives the male gamete.

CARPOSPORE. Spore produced after sexual reproduction in red Algae; borne at end of outgrowth of fertilized carpogonium. See *Cystocarp*.

CARPUS. Region of fore-leg of tetrapod vertebrates containing carpal bones (q.v.); roughly the wrist.

CARTILAGE. Skeletal tissue of vertebrates consisting of rounded cells scattered in a resilient polysaccharide-containing matrix, with numerous collagen fibres. Devoid of blood-vessels in adult. A cartilage-like substance occurs in Cephalopoda (squids, octopuses), where it supports the brain.

CARTILAGE-BONE (REPLACING BONE). A bone which replaces embryonic cartilage, e.g. any limb bone; hip girdle; vertebral column; parts of skull such as auditory capsule. Cf. *Dermal bone*.

CARUNCLE. Warty outgrowth seen on seeds of a few flowering plants, e.g. castor oil, obscuring micropyle (q.v.).

CARYOPSIS. Fruit of grasses. An achene (q.v.) with ovary wall (pericarp) united with seed coat (testa).

CASPARIAN STRIP. Area impervious to water in form of a strip running completely round radial and transverse walls of cells of endodermis (q.v.), resulting from impregnation of primary wall with substances giving reaction for suberin and lignin, respectively; may be as wide as wall itself, or narrow, thread-like. Thought to ensure that movement of water and solutes across endodermis is completely subject to regulatory activity of living protoplasm of its cells.

CASTE. Of social insects, a type of structurally and functionally specialized individual. E.g. among hive bees there are three castes, queens (fertile females), workers (females, usually sterile), drones (males). Among ants there may be several different kinds of workers (all sterile females). In termites the distinction between castes is not related to sex (as it is in the bees and ants), e.g. 'soldiers' may be sterile males or females. See *Polymorphism.*

CATABOLISM (KATABOLISM). Breaking down by living things of complex organic molecules, with liberation of energy. See *Metabolism*, *Anabolism.*

CATALASE. Enzyme which breaks down the poisonous substance, hydrogen peroxide, formed during plant and animal metabolism, to water and oxygen. Its prosthetic group (q.v.), containing iron, is same as that of haemoglobin.

CATARRHINE. Old world anthropoid. Characterized by narrow nasal septum and by menstrual cycle, e.g. baboon, chimpanzee, man. Cf. *Platyrrhine.*

CATERPILLAR. Larva of certain insects (e.g. butterflies and moths); soft bodied, worm-like, with three pairs of jointed legs on thorax, and short unjointed prolegs on abdomen.

CATHEPSIN. System of intracellular proteolytic enzymes responsible for autolysis.

CATKIN. Kind of inflorescence (q.v.).

CAUDAD. Towards the tail.

CAUDAL. Concerning the tail.

CAULINE. Belonging to the stem or arising from it.

C.C. Cubic centimetre.

CELL. (Zool.). Discrete mass of protoplasm, bounded by a plasma membrane. (Bot.). As in Zoology, but including also the surrounding cell wall (usually of cellulose or chitin). Sometimes used for cell-wall only.

Most animals and plants can conveniently be considered as made up of two components, though of course these form an integrated system in the normal organism: (*a*) the protoplasm, in which are situated the systems of enzymes (q.v.) responsible for controlling the chemical reactions (metabolism, q.v.) characteristic of living things, and the nucleic acid systems responsible for synthesis of the enzymes and of themselves; (*b*) a component, made *by* the protoplasm, devoid of these enzyme systems, largely composed of fibrous materials which provide mechanical support for the protoplasm, e.g. cellulose in plants, collagen in animals.

The protoplasmic component of the larger plants or animals is not usually a continuous mass; it consists of numerous discrete units. Each unit is microscopically small (a man contains something like a million million of them) each is bounded by an extremely thin membrane (see *Plasma-membrane*) which allows some substances to pass through but not others, and so partially isolates the inside of each unit from its surroundings; and usually each contains one nucleus (q.v.), sometimes more than one, very rarely none. The shapes of such units are very various, and they may often have some special function highly developed, e.g. power of contraction in muscle; of photosynthesis in green plants. But all of them have the common features of metabolism (q.v.). Every unit originates from a pre-existing unit, usually by division into two, involving mitosis (q.v.) or some similar process which distributes the nuclear material between the two products (but see also *Fertilization*).

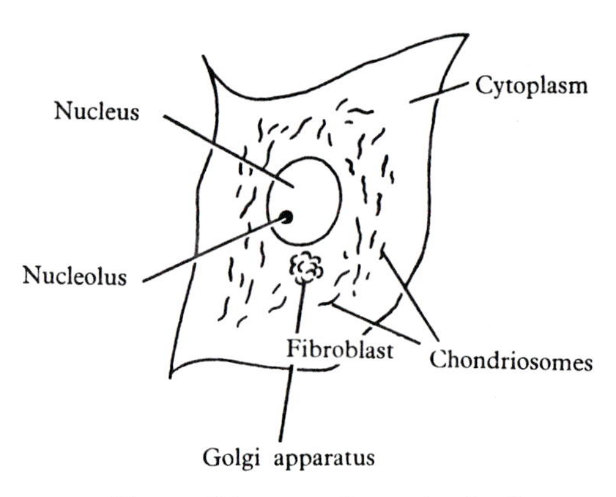

Fig. 1. Diagram of an animal cell.

Each of these units in an animal is called a cell. Each in a plant is usually called a protoplast, every protoplast being consistently surrounded, with peculiar intimacy, by a wall commonly of cellulose (which a zoologist would regard as intercellular material). The protoplast and its cell-wall are together known as the cell in botany. Historically the first use of the word cell was for the plant cell-wall and the space it enclosed: and even when the space contains no protoplast, the word is still applied in this way. Since there is rarely a parallel in animals to the plant cell-wall, divergent usages have unfortunately developed.

Although a cell or protoplast has been characterized above as a discrete unit bounded by a membrane, there are cases where, while the elements are almost entirely separated from each other by their membranes, they are in places continuous (by intercellular bridges or plasmodesmata, q.v.). Each almost discrete element is still called a cell (animal) or protoplast (plant). There exist also large continuous masses of protoplasm with many nuclei, which are regarded as units and called syncytia, plasmodia or coenocytes rather than cells or protoplasts. There is no agreed demarcation between a collection of connected cells and a syncytium, etc. Many microscopic organisms (Bacteria, Algae, Fungi, Protozoa) consist of a single cell throughout life, in plants and bacteria complete with cell-wall, each with the characteristics (listed above) of a plant or animal protoplasmic element, though naturally without specialization of particular function, since all functions must be fulfilled by the one cell. Every multicellular animal or plant which reproduces sexually, as most do, starts as a single cell (formed by fusion of a male cell with a female cell), which by continuous doubling forms the numerous cells of the adult.

CELL-BODY (PERIKARYON). The mass of cytoplasm with contained nucleus, from which arise the branches of a nerve-cell (q.v.).

CELL-DIVISION. Division of cell, both cytoplasm and nucleus, into two. The nucleus usually divides by mitosis (q.v.), occasionally by amitosis (q.v.). The cytoplasm divides by constriction into two in animals; by formation of a middle lamella across its equator in plants.

CELL-LINEAGE. Developmental history in terms of descent by cell division of later cells from earlier cells. The cell-lineage of an organ traces the succession of cells, from the zygote onwards, which culminates in the group of cells constituting that organ.

CELL-MEMBRANE. Plasma-membrane (q.v.).

CELL-SAP. Complex solution of inorganic and organic substances contained in the vacuole of a plant cell.

CELL-THEORY. Theory, initiated by Schleiden and Schwann, 1838–9, that all animals and plants are made up of cells (q.v.) and their products, and that growth and reproduction are fundamentally due to division of cells.

CELLULOSE. Fundamental constituent of cell wall in all green plants and in a few Fungi. Complex polysaccharide carbohydrate formed by condensation of many glucose molecules; with a fibrous structure to which is due its use in textile industries (cotton, linen, artificial silk).

CELL-WALL. Limiting layer of plant cells, formed by, and closely investing, the protoplasm; comparatively rigid and giving mechanical support to plant tissue. In living cells traversed by extremely fine cytoplasmic threads, the plasmodesmata, which form delicate protoplasmic connexions between adjacent cells. The walls of newly formed cells are at first very thin but as the cells assume their permanent character the walls thicken. In an older cell one can distinguish primary and secondary wall layers. The primary layer (middle lamella) represents the fused primary walls of adjacent cells; it consists of pectic substances. On this, the secondary layers, mainly of cellulose, are deposited. During the deposition of these secondary layers, certain small areas remain unthickened, forming depressions (pits). Pits of adjacent walls usually coincide so that in these areas the protoplasts are separated only by the middle lamella, which forms the pit membranes, through which pass the majority of the plasmodesmata. Some cell walls undergo further modification, e.g. cuticularization of epidermal cells and suberization of cork cells, rendering them impermeable to water; lignification of fibres, vessels, tracheids, conferring on them increased strength and rigidity.

CEMENT. Bone-like substance which makes a thin covering to root of vertebrate tooth (i.e. below gum level) and in some mammals (e.g. ungulates) covers parts of enamel of crown.

CENOZOIC. See *Geological period.* The present era, the age of mammals; extends from about 70 million years ago.

CENTRAL NERVOUS SYSTEM (C.N.S.). A mass of nervous tissue which co-ordinates the activities of an animal. In vertebrates the C.N.S. is the brain and spinal cord, which together are basically a single hollow tube with very thick walls, enclosed within skull and backbone. In most invertebrates the C.N.S. consists of a few solid cords of nervous tissue, often associated with larger masses called *ganglia* (q.v.). In some invertebrates however (Coelenterata, Echinodermata) the place of a C.N.S. in providing co-ordination is taken by a diffuse *nerve-net* (q.v.).

The C.N.S. is in a position to co-ordinate the activities of an animal with each other and with the environment because almost all the messages (impulses, q.v.) sent from the sense-organs travel straight into the C.N.S., and almost all impulses received by the muscles, glands, etc., come out from the C.N.S. The C.N.S. contains many fairly direct nervous pathways between certain sense-organs and certain muscles or glands (see *Reflex*) which ensure that some stimuli are followed rapidly by their appropriate standardized responses. But there is also great opportunity in the C.N.S. for co-ordination of a variety of stimuli by interaction between nerve cells, because of the large number of junctions (synapses) between nerve cells which occur there. In vertebrates for instance, the great majority of synapses occurs in the C.N.S. The C.N.S., however, is not to be regarded simply as a means of reconciling and co-ordinating the immediate activities of the sense-organs; it has an activity of its own independent of the stimuli of the moment, most obvious in the influence of past experience on present performance. The rest of the nervous system, other than the C.N.S., is the peripheral nervous system. See *Nervous System, Nerve-cell.*

CENTRE (NERVE). Region of central nervous system with a restricted special function, e.g. respiratory centre in medulla of vertebrates which controls respiratory movements. The term is physiological (cf. *Nucleus of Brain*).

CENTRIOLE. Minute granule, present in many resting cells, just outside the nuclear membrane. Doubles before mitosis, and at mitosis the two centrioles move apart and form the poles of the spindle, and the centres of the asters when present. Absent in higher plants. Centriole may also be connected to axial filaments of flagella and cilia, e.g. axial filament of sperm tail grows out from centriole.

CENTROMERE. Spindle-attachment (q.v.).

CENTROSOME. Region of differentiated cytoplasm containing centriole.

CENTRUM. Massive part of each vertebra, lying ventral to spinal cord. Replaces embryonic notochord. Each is firmly but flexibly attached to adjacent centra by collagen fibres.

CEPHALIC INDEX. Measure of shape of human head: breadth as percentage of length (back to front).

CEPHALOCHORDATA. Acrania (q.v.).

CEPHALOPODA (SIPHONOPODA). Class of Mollusca including octopus, squids, cuttle-fish. Head well developed, with crown of mobile tentacles; eyes large and complex; nervous system highly centralized.

CERCARIA. One of the larval forms of flukes (Trematoda); produced asexually by redia larvae (q.v.) while these are parasitic in snails. Cercariae are infective to a new host to which they gain access either in food (sheep liver-fluke) or by penetrating skin (*Schistosoma*). This new host may be the definitive one, in which the cercaraie become sexually mature flukes; or in some species it may be a second intermediate host, where they live for some time before being swallowed by definitive host and becoming sexually mature.

CERCI. Paired appendages, often sensory, at hind end of abdomen of many insects. 'Forceps' of earwigs are modified cerci.

CEREAL. Flowering plant of the family Gramineae whose seeds are used as food, e.g. wheat, oats, barley, rye, maize, etc.

CEREBELLUM. Outgrowth of anterior end of dorsal surface of hind-brain of vertebrates, varying from a mere transverse commissure in cyclostomes to a conspicuous part of the brain in birds and mammals (with a cortex of grey matter in mammals). Particularly concerned in co-ordination of complex muscular movements.

CEREBRAL CORTEX. Layer of grey matter (q.v.) rich in synapses, superficial to white matter, present in cerebral hemispheres of amniotes, and some anamniotes, but really extensive only in mammals (neopallium, q.v.).

CEREBRAL HEMISPHERES. Paired outpushings of front end of fore-brain of vertebrates. In early vertebrates mainly concerned with sense of smell; but in amniotes general co-ordinating functions become predominant; and in mammals hemispheres are largest part of brain, with a much folded cortex by which much of animal's activities are controlled.

CEREBROSPINAL. (Adj.). Of the brain and spinal cord of vertebrates.

CEREBROSPINAL FLUID (C.S.F.). Fluid which fills the cavity inside vertebrate brain and spinal cord, and the pia-arachnoid (q.v.) space outside them, the inside and outside communicating by holes in the roof of the hind-brain. Secreted continuously by choroid plexuses (q.v.) and reabsorbed by veins of surface of brain. It is a solution of the small molecules of the blood (salt, glucose, etc.) almost devoid of protein and containing very few cells. About 100 ml. is present in man.

CERUMEN. Ear wax. Secretion of skin glands (perhaps modified sweat glands) of external ear passage in mammals.

CERVICAL. (Adj.). Of the neck. *C. vertebrae* have reduced or absent ribs; they are concerned in movement of head. Mammals (even giraffes) are peculiar in having seven, whatever the length of their neck, with the exception of the manatee (*Manatus*, order Sirenia, 6) and two kinds of sloth (*Bradypus*, 9 or 10, and *Choloepus*, 6, order Xenarthra).

CERVIX (C. UTERI). Cylindrical posterior part of the uterus of mammals, which leads into the vagina. Contains numerous glands supplying mucus to vagina.

CESTODA. Tapeworms (q.v.). Class of Platyhelminthes. Parasites; with cuticle; without gut. Co.nplicated life cycle.

CETACEA. Whales, porpoises, dolphins. An order of placental Mammals. They are completely aquatic, and have many special features connected with this mode of life, such as layer of fat (blubber) in the skin; no hind limbs and only traces of pelvic girdle; fore-limbs paddle-like; tail-fin transverse (unlike fishes); often a dorsal fin.

CHAETA. Bristle, made largely of chitin, characteristic of chaetopod Annelida (bristleworms, earthworms) in which they occur in groups projecting from the skin, segmentally arranged. In some cases assist locomotion, e g earthworm. Term may also be applied to bristles of other animals, e.g. of Brachiopoda. See *Seta*.

CHAETOGNATHA. Arrow-worms. Small phylum of marine animals of doubtful affinities. Includes *Sagitta*, abundant in plankton.

CHAETOPODA. A class of Annelida including bristleworms (order Polychaeta) and earthworms (order Oligochaeta). Well-segmented; coelom spacious; chaetae present.

CHALAZA. Basal region of Angiosperm ovule where stalk (funicle) is attached.

CHAMAEOPHYTES. Class of Raunkiaer's Life Forms (q.v.).

CHELA. A pincer, e.g. of lobster.

CHELONIA. Tortoises and turtles. Order of reptiles with body enclosed in plates of bone usually covered by epidermal horny plates (tortoiseshell). Shoulder and pelvic girdles unique in being inside the ribs. No teeth.

CHEMOAUTOTROPHIC (CHEMOSYNTHETIC). Obtaining energy from a simple inorganic reaction, the nature of which varies according to species, e.g. oxidation of hydrogen sulphide to sulphur by *Thiobacillus*. Present in several kinds of autotrophic bacteria.

CHEMORECEPTOR. Receptor which detects and differentiates substances according to their chemical structure, by contact with their molecules, e.g. smell, taste.

CHEMOSYNTHETIC. Chemoautotrophic (q.v.).

CHEMOTAXIS. Taxis (q.v.) in which stimulus has the form of a gradient of chemical concentration, e.g. movement of polymorph leucocytes towards bacteria.

CHEMOTROPHIC. Of organisms, obtaining energy by chemical reactions independent of light. The reactions may be based on inorganic or organic substances obtained from the environment. Cf. *Autotrophic*, *Heterotrophic*, *Phototrophic*.

CHEMOTROPISM. (1) (Bot.). Tropism (q.v.) in which stimulus has the form of a gradient of chemical concentration, e.g. downward growth of pollen tubes into stigma due to presence of sugars. (2) (Zool.). Often used synonymously with chemotaxis (q.v.).

CHIASMA (PL. CHIASMATA). (1). An interchange at corresponding points between homologous chromosomes; the basis of crossing-over (q.v.) See *Meiosis*. Fig. 6B, page 148. (2). See *Optic Chiasma*.

CHILOPODA. Centipedes. Order (or class) of Arthropoda. See *Myriapoda*.

CHIMAERA. (1) Organism whose tissues are of two or more genetically different kinds. Can occur as a result of mutation or abnormal distribution of chromosomes, affecting one cell and all its descendants, during development; or as a result of natural or artificial grafting of two different plants, whose cells become associated, producing mixing of characters: such a plant is often called a *graft hybrid*. See *Mosaic*. (2) Generic name of a fish 'King of the Herrings' (Holocephali).

CHIROPTERA. Bats. Order of placental mammals. Characterized by membranous wing spread between arms, legs, and sometimes tail, and supported by greatly elongated fingers.

CHITIN. Nitrogen-containing polysaccharide (q.v.) with long fibrous molecules, forming material of considerable mechanical strength and resistance to chemicals. Present in covering layer (cuticle, q.v.) of insects. Similar substance is found in cell walls of many fungi.

CHLAMYDOSPORE. Thick-walled fungus spore capable of surviving conditions unfavourable to growth of the fungus as a whole; asexually produced from a cell or portion of a hypha.

CHLORENCHYMA. Parenchymatous tissue containing chloroplasts (q.v.).

CHLOROCRUORIN. Respiratory pigment (green, fluorescing red) in blood of certain Polychaeta. A protein containing iron; closely related to haemoglobin.

CHLOROPHYCEAE. Green algae. Class of Algae with chlorophyll located in well-defined chloroplasts, not masked by other pigments as in other classes of the group. Widely distributed, aquatic, both in fresh and salt water, or terrestrial in damp places, e.g. soil, shaded walls, tree trunks, etc. Primitive forms microscopic, unicellular, motile by flagella, or non-motile; occurring singly or grouped into colonies. Higher forms multicellular with filamentous or flattened thallus. Asexual reproduction by cell division, fragmentation of thallus or by zoospores. Sexual reproduction isogamous or anisogamous, both gametes, or only male gamete, motile.

CHLOROPHYLL. Green pigment found in all members of the plant kingdom except Fungi and a few flowering plants; a closely related substance, *bacteriochlorophyll*, occurs in a few autotrophic (q.v.) Bacteria. Localized within cells in bodies known as chloroplasts, except in blue-green algae (Cyanophyceae) where it is distributed throughout cytoplasm. Consists of a mixture of two pigments, chlorophyll a (formula $C_{55}\ H_{72}\ O_5\ N_4\ Mg$), and chlorophyll b (formula $C_{55}\ H_{70}\ O_6\ N_4\ Mg$), chemically related to prosthetic groups of haemoglobin and cytochrome. By means of chlorophyll green plants build up carbohydrates (q.v.) from carbon dioxide and water, absorbing energy from sunlight, in the process known as *photosynthesis*.

CHLOROPLAST. Plastid (q.v.) containing chlorophyll; occurring in numbers in cells of leaves and young stems and responsible for green colour of plants.

CHLOROSIS. Disease of green plants characterized by yellow (chlorotic) condition of parts that are normally green; caused by conditions preventing chlorophyll formation, e.g. lack of light.

CHOANAE (INTERNAL NARES). Of vertebrates. Internal openings of nasal cavity (q.v.) into mouth. Present, as are also external nares opening on to face, in group Choanata; remaining vertebrate classes have only external nares. Situated near front of roof of mouth, except in those choanates with a false palate (see *Palate*) in which they are at the back.

CHOANATA. Grouping of vertebrate classes sometimes used in classification. Includes Choanichthyes, Amphibia, Reptilia, Aves (birds), and Mammalia. These are more closely related to each other than to the other vertebrate classes. Have axial skeleton in paired limbs, and nostrils opening into the mouth (choanae) as well as on to face.

CHOANICHTHYES. Class of fish (sometimes regarded as a sub-class of class Osteichthyes) containing Crossopterygii and Dipnoi. Characterized by bony skeleton; paired fins with a central skeletal axis; nostrils opening both on to face and into mouth; cosmoid scales (q.v.) or derivatives of them. Cf. *Actinopterygii* and *Chondrichthyes*, the other two classes of living fish.

CHOANOCYTE (COLLAR-CELL). Peculiar kind of cell with a flagellum surrounded by a thin protoplasmic sheath or collar, found only in sponges and a small group of Flagellata (Choanoflagellata).

CHOLESTEROL. A sterol found in all animals studied, but not in plants. A vitamin for insects, not for vertebrates.

CHOLINE. An organic base, a vitamin for cockroaches, and probably for some mammals which do not synthesize all they require. Choline is a constituent of certain important fats (lecithin, q.v.) and of acetylcholine (q.v.). Formula: $HO.C_2H_4.\ N(CH_3)_3.\ OH$.

CHOLINERGIC. Of a nerve fibre, secreting acetylcholine (q.v.) at its end when nerve impulses arrive there. In vertebrates, motor fibres to striped muscle, parasympathetic fibres to smooth muscle, and fibres connecting C.N.S. to sympathetic ganglia are all cholinergic. Cholinergic motor fibres occur in Annelids. Cf. *Adrenergic*.

CHOLINESTERASE. Enzyme which splits and hence inactivates acetylcholine (q.v.).

CHONDRICHTHYES. Cartilaginous fish. Class of vertebrates containing the Selachii (sharks, dogfish, skates, rays), Holocephali (chimaeras), and some fossil groups. Characterized by absence of true bone (although the cartilaginous skeleton is often calcified); complicated copulatory organs ('claspers') formed from pelvic fins of males; denticles (q.v.). (Cf. *Actinopterygii* and *Choanichthyes*, the other two classes of living fish.) Almost exclusively marine, and unlike the other fish classes, have been so throughout their known fossil history, though ancestors probably lived in fresh-water. First appear in late Devonian. See *Elasmobranchii*.

CHONDRIOSOMES. Mitochondria (q.v.).

CHONDROCRANIUM. The part of skull first formed in embryo of vertebrates, as cartilaginous investment of part of brain and inner ear. In most vertebrates some or all of this cartilage is later replaced by bones, and the skull completed by membrane bones external to chondrocranium.

CHORDA-MESODERM. Mesoderm and notochord of vertebrate embryo. In early stages of development, before differentiation, these tissues often form a continuous mass or layer of similar cells, and they are closely related in physiology of development also, e.g. see *Organizer*; consequently it is convenient for some purposes to group them together.

CHORDATA. Phylum of animals with notochord (q.v.), hollow dorsal nerve cord, and gill slits. Includes subphyla Vertebrata (Craniata) and Protochordata.

CHORIO-ALLANTOIC GRAFTING. Grafting living pieces of tissue on the outer surface of chorion, in region where allantois is fused with chorion, of a young living chick embryo. After the shell is sealed, chick continues development and blood-vessels of allantois invade graft and maintain it.

CHORION. (Zool.). Embryonic membrane of amniote vertebrate, consisting of outer ectodermal epithelium with layer of mesoderm beneath. In Sauropsida, the more superficial of the two layers which enclose the amniotic cavity. See *Amnion*. In mammals, the superficial layer enclosing all the embryonic structures (equivalent therefore to sauropsid chorion plus outer layer of yolk-sac); its outer epithelium is trophoblast (q.v.); forms placenta (q.v.). See Fig. 8, p. 184. (2) Non-cellular membrane secreted around ovum by cells of ovary, e.g. the superficial envelope or 'shell' of an insect egg.

CHOROID. Layer, in vertebrate eye, immediately outside retina, containing blood-vessels and pigment. See Fig. 3, p. 86.

CHOROID PLEXUS (CHORIOID PLEXUS). Numerous projections, into cavity (ventricle) of brain, of the non-nervous epithelium which constitutes its roof in some regions, enclosing tufts of blood-vessels. Secretes cerebrospinal fluid. One plexus occurs in roof of each of the four ventricles in man.

CHROMATID. One of the two strands, which result from duplication of a chromosome, found during prophase and metaphase of mitosis or meiosis. They separate at anaphase and are then known as daughter-chromosomes.

CHROMATIN. Nucleo-protein of chromosomes, which stains strongly with basic dyes.

CHROMATOPHORE. (1) (Bot.). Chromoplast (q.v.). (2) (Zool.). Cell with pigment in its cytoplasm. In many vertebrates (e.g. chamaeleon, frog) and Crustacea, rapidly changing disposition (concentration or dispersion) of the pigment within the cell, alters colour of animal.

CHROMOMERES. Darkly staining granules found in constant arrangement along prophase chromosomes during meiosis. Chromomeres in corresponding positions on homologous chromosomes pair during meiosis. See *Salivary Gland Chromosomes*.

CHROMOPLAST (CHROMATOPHORE). Pigmented plastid (q.v.) of plant cells. May be red, orange, or yellow, containing carotin and/or xanthophyll; or green, containing chlorophyll. Former are common in fruits and flowers; the latter (chloroplasts, q.v.) are responsible for the green colour of plants. In Algae variously coloured owing to presence with chlorophyll of other pigments, xanthophyll, carotene, fucoxanthin, phycoerythrin, phycocyanin.

CHROMOSOME. Thread-shaped body, numbers of which occur in the nucleus of every animal or plant cell. Bacteria and viruses have structures of similar function, though somewhat different composition. They occur in pairs, usually several different pairs per nucleus, in somatic cells of animals and higher plants (but see *Endomitosis*). The two members of each pair are identical in appearance and are said to be *homologous* (to each other). Homologous chromosomes associate in a characteristic way (pairing, q.v.) during meiosis. Chromosomes of different pairs are often visibly different from each other in e.g. size, shape. There are one to over 100 pairs per nucleus according to species (man has twenty-three, *Drosophila melanogaster* four pairs) and most cells of most individuals of a given species have a set of similar chromosomes which is characteristic of that species. Gametes, and cells of the gametophyte (q.v.) of plants, have only one member of each pair in their nuclei. Chromosomes are usually visible only during mitosis or meiosis, when they contract in length, finally becoming short thick rods (a few microns long), probably by coiling up into a spiral. Fine threads have however been seen in the living resting nucleus which are probably the chromosomes. Chromosomes consist largely of nucleo-protein (q.v.), the nucleic acid being DNA, and they stain strongly with basic dyes (they are *basophilic*) during mitosis and meiosis.

Chromosomes are elaborately differentiated in structure along their length. This differentiation consists of a series of different genes (q.v.), according to genetic theory, in single file, perhaps with other material between them; a spindle-attachment at some point in their length; and regions of heterochromatin (q.v.). The arrangement of these elements is practically the same in all homologous chromosomes. In some species the arrangement of many of the genes (i.e. of the loci) has been discovered (see *Chromosome Map*); it can be correlated with visible markings on the giant salivary gland chromosomes (q.v.).

Before every nuclear division (mitosis, q.v.) each chromosome doubles and the two duplicates separate to the two daughter nuclei. For the behaviour of chromosomes during gamete formation, of great importance to genetics, see *Meiosis*, Fig. 6B, p. 148.

CHROMOSOME MAP. Plan showing positions of the genes on a chromosome. It can be constructed from data of crossing-over (q.v.), the frequency of crossing-over between the genes taken in pairs indicating their linear order, i.e. *relative* positions; the distances on such a map are given in units of cross-over value. Or it can be constructed by observations of localized aberrations of a chromosome (especially salivary) which produce phenotypes known to be associated with presence or absence of particular genes. The results of the two methods agree.

CHRYSALIS. Pupa (q.v.).

CILIARY BODY. Thickened circular rim of choroid of vertebrate eye, at border of cornea, containing *ciliary muscles* which produce accommodation (q.v.). To it the lens is attached; and from it the iris springs. See Fig. 3, p. 86.

CILIARY FEEDING. Feeding by filtering minute organisms from a current of water drawn through or towards the animal by cilia, as in amphioxus.

CILIATA. Group of ciliated unicellular organisms. See *Ciliophora.* Includes *Paramecium.*

CILIATED EPITHELIUM. (Zool.). Sheet of cells with cilia on exposed surface. The cilia beat in co-ordinated rhythm. A common method of moving fluids in animal body, e.g. in many microphagous animals (q.v.) or in respiratory passages of land vertebrates.

CILIOPHORA. Class of Protozoa, characterized by the possession of cilia and double nucleus (macro- and micro-nucleus). Some of the most complex single-celled organisms known belong to this group. Includes Ciliata, which are ciliated throughout life, and Suctoria, which are ciliated only in their immature form.

CILIUM (PL. CILIA). Fine cytoplasmic thread projecting, along with many others, from the surface of a cell. Cilia of a cell or of a whole epithelium lash with orderly beat in a constant direction. Each cilium moves the fluid surrounding it by lashing stiffly through it like an oar, and then bending on its recovery stroke so as to offer less resistance. Have same internal structure as flagellum (q.v.) from which they probably only differ in their relative shortness. Present in one group of Protozoa (Ciliophora) and all Metozoa except Arthropoda and Nematoda. In Metazoa commonly occur on *ciliated epithelia.* Cilia like those of animals occur only in few plants, e.g. Cycads; but the term is often used in botany synonymously with flagellum (q.v.).

CIRCULATORY SYSTEM. See *Blood System.*

CIRCUMNUTATION. See *Nutation.*

CIRREPEDIA. Barnacles. Sub-class of Crustacea; aquatic; sessile. Feed by drawing a current of water towards the mouth with their whip-like appendages; or parasitic.

CITRIC ACID CYCLE. Krebs Cycle (q.v.).

CLADOCERA. Water-fleas. Order of Crustacea.

CLADODE (PHYLLOCLADE). Modified stem having appearance and function of a leaf, e.g. butcher's broom.

CLASS. One of the kinds of group used in classification of organisms, e.g. Mammalia; Dicotyledoneae. Consists of a number of similar orders (sometimes of only one order). Similar classes are grouped in a phylum (or more usually in Botany in a division). See *Classfication.*

CLASSIFICATION. Organisms are classified scientifically in a hierarchical series of groups. The smallest group regularly used is the species (q.v.). Species which are more like each other than like other species are grouped together in a genus; similarly genera are grouped into families, families into orders, orders into classes, and classes into phyla or divisions. The term Animal and Plant Kingdom are also used; and various sub-groups according to convenience. All groups of any one kind, e.g. all families, are supposed to have approximately the same 'weight', e.g. every family is supposed to differ from its related families belonging to the same order, by a roughly equal amount and the same degree of difference should be found in every order. Consequently a single peculiar family like that comprising genera of amphioxus cannot be put in an order along with any other family, and must form an order, and in this case a class and sub-phylum, by itself.

Since the more closely related by descent organisms are, the more features they have in common, biological classification reflects, and is usually intended to reflect, degrees of evolutionary relationship. The system is not a rigid one. Biologists differ among themselves on many points of classification, and the system is always subject to modification as more and more is learnt about organisms, and about their fossil history. At present the criteria used for classifying animals and plants are almost entirely structural; but as we learn more of biochemical and physiological differences, other criteria will no doubt be used in constructing the classificatory system.

CLAVICLE. Dermal bone of ventral side of shoulder-girdle of many vertebrates. Collar-bone of man. See Fig. 10, p. 216.

CLEARING. Part of usual process of preparation of tissues for microscopical examination. Consists in soaking tissue in substances such as benzene or xylene, after dehydration (q.v.). These substances bridge the gap, caused by mutual insolubility, between the alcohol used in dehydration and the substances usually used in the next stage of preparation, e.g. paraffin wax (see *Microtome*), or Canada balsam (q.v.), by being miscible with both. Incidentally the tissue becomes very transparent in most clearing agents.

CLEAVAGE (SEGMENTATION). Repeated subdivision of the zygote cytoplasm, accompanying a corresponding nuclear mitosis, which follows fertilization. In animals often produces a mass of small cells, the blastula. Usually called segmentation in plants. See *Holoblastic*, *Meroblastic*.

CLEIDOIC EGG. Egg of terrestrial animal enclosed within protective shell, which largely isolates it from its surroundings, permitting only respiration and some loss (occasionally gain) of water. See *Uricotelic*. E.g. egg of bird or insect. Contrasted with most marine eggs, which exchange water and salts fairly freely with their surroundings.

CLEISTOCARP. Completely closed fruit body of certain Ascomycete Fungi, e.g. powdery mildews, from which spores are liberated by decay or rupture of its wall.

CLEISTOGAMY. Fertilization within an unopened flower, e.g. violet.

CLIMAX. Type of plant community the composition of which is more or less stable, in equilibrium with existing natural environmental conditions, e.g. oak forest in lowland Britain. Cf. *Plagioclimax*.

CLINE. Continuous gradation of form differences in a population of a species, correlated with its geographical or ecological distribution.

CLISERE. Succession of climaxes (q.v.) in an area as a result of climatic changes.

CLITELLUM. Saddle-like region of some Annelida (earthworms, leeches). Prominent in sexually mature animals. Contains glands which secrete a mucus sheath around copulating animals thus binding them temporarily together, and a cocoon around fertilized eggs.

CLITORIS. Homologue of penis in female mammal.

CLOACA. Terminal part of gut of most vertebrates into which kidney and reproductive ducts open; in such cases there is only one posterior opening to the body, the cloacal aperture, instead of (as in most mammals) separate anus and urinogenital openings. Also terminal part of intestine in some invertebrates, e.g. sea cucumbers.

CLONE. The descendants produced vegetatively or by apomixis from a single plant; asexually or by parthenogenesis from a single animal. The members of a clone are of the same genetic constitution, except insofar as mutation occurs amongst them.

CLUB MOSS. See *Lycopodiales*.

CNIDARIA. Hydroids, jellyfish, sea-anemones, corals. Subphylum of Coelenterata. With thread cells; body a polyp or medusa; locomotion by muscular action, not cilia (cf. *Ctenophora*). Contains classes Hydrozoa, Scyphozoa, Actinozoa.

CNIDOBLAST. Thread cell (q.v.).

CNIDOCIL. Sensory hair of thread cell (q.v.).

C.N.S. Central nervous system (q.v.).

COBALAMINE (VITAMIN B_{12}). Cobalt-containing vitamin required by many organisms. Lack of it upsets cell division. In man, the haemopoietic principle necessary for red blood cell formation is produced from the vitamin by a substance secreted in the stomach.

COCCIDIOSIS. Disease of rabbits, fowls, etc., caused by certain Sporozoa parasitic in, e.g. intestine, liver.

COCCUS. Globular bacterium.

COCCYX. Tail vertebrae fused together. In man, consists of three to five vestigial vertebrae.

COCHLEA. That part of the inner ear concerned in the reception of sound with analysis of its pitch. A projection of the saccule. Found in crocodiles, birds, and mammals. In the latter coiled in a spiral.

COCOON. Protective covering of eggs, larvae, etc.; e.g. eggs of some annelids are fertilized and develop in a cocoon. See *Clitellum*. Larvae of many insects spin a cocoon in which pupae develop (cocoon of silkworm moth is source of silk).

CO-DOMINANT. See *Dominant*.

COELACANTHINI. A large order of Crossopterygii, almost all fossil (from Devonian onwards) and fresh-water, but with living marine representatives off S. Africa (see *Latimeria*).

COELENTERATA. Phylum of animals containing hydroids, jellyfish, sea-anemones, corals, comb-jellies. All aquatic, most marine. Body built on a fairly simple plan; more or less radially symmetrical; of jelly-like consistency; gut (coelenteron) has one opening only;

nervous system diffuse; no excretory system; no blood system. The body-wall is usually described as having only two layers of cells (dipoblastic, q.v.); but while this statement is true of the delicate hydroids and of *Hydra* which is usually cited as a 'type' of the group, it needs qualification for jellyfish, sea-anemones, and comb-jellies. See *Mesogloea.* Phylum contains subphyla Cnidaria (classes Hydrozoa, Scyphozoa, Actinozoa) and Ctenophora.

COELENTERON. The single cavity within the body of a coelenterate. It serves as a gut, and in some groups eggs and sperm are discharged into it. Has a single opening.

COELOM. Main body cavity (q.v.) of many triploblastic animals, in which gut is suspended. Situated in the mesoderm, lined by epithelium (mesothelium). Contains fluid which, unlike blood, is not circulated by muscular walls (except in some leeches). Germ-cells mature in its walls or in a specialized part of it (the gonads); and when they are ripe they are often shed into the coelom, and are then usually transported to exterior by *coelomoducts* (e.g. oviducts of vertebrates). In many animals plays important part in collecting excretions which are removed from it by nephridia or coelomoducts. Coelom is spacious in Echinodermata, Vertebrata, polychaete and oligochaete Annelida (though not in Hirudinea) where it forms the perivisceral cavity. In Arthropoda and Mollusca it forms only the cavity of gonads and excretory organs, and not the perivisceral cavity, which is a haemocoel.

COELOMODUCT. Duct formed from lining of coelom (i.e. from mesoderm) connecting coelom with exterior; sometimes used for conveying products of gonads to exterior; sometimes excretory, e.g. kidneys of Mollusca.

COENOBIUM. Type of colony found in certain groups of Algae; consisting of a definite number of cells arranged in a specific manner, e.g. in a hollow sphere as in *Volvox.*

COENOCYTE. (Bot.). Multinucleate mass of protoplasm formed by division of nucleus but not of cytoplasm of an original cell with single nucleus, e.g. many Fungi, especially Phycomycetes and some green Algae. Cf. *Plasmodium, Syncytium, Cell.*

COENOSPECIES. Group of species related by the possibility, directly or indirectly, of hybridization one with another.

COENZYME. Organic compound playing essential part in reaction catalysed by some enzyme or system of enzymes, without being consumed in the process, usually acting as temporary carrier of intermediate product of the reaction. Many coenzymes are known, and the same one may function in reactions catalysed by many different enzymes (e.g. adenosine triphosphate, ATP, q.v.).

COLCHICINE. Drug (an alkaloid) which prevents mitosis proceeding beyond metaphase (by inhibiting spindle). In actively dividing tissue of an organism treated with colchicine there is an accumulation of cells arrested in metaphase making easy the detection of pro liferating regions. Reversion of colchicine-inhibited metaphase nuclei to the resting state results in polyploidy (q.v.) since the chromosomes have duplicated without the chromatids separating into daughter nuclei. This is a method of inducing polyploidy artificially, important in obtaining new agricultural and horticultural varieties. See *Allotetraploid.*

COLEOPTERA. Beetles. A very large order of endopterygote insects. Forewings horny, covering hind part of body; hind wings membranous, may be small or absent. Larvae may be active predators (campodeiform) or caterpillar-like (eruciform) or a grub (apodous). Some important agricultural pests belong to this order, e.g. bollweevil, wireworm.

COLEOPTILE. Protective sheath surrounding plumule (q.v.) in grass seedlings.

COLEORHIZA. Protective sheath surrounding radicle (q.v.) in grass seedlings.

COLLAGEN. Fibrous protein, which on boiling yields gelatin. Forms intercellular fibres ('white fibres' of vertebrate connective tissue). One of principal skeletal substances, binding cells and tissues together, in vertebrates and probably in most Metazoa. Fibres made of collagen have a high tensile strength (exemplified by tendon) but unlike elastin fibres have little reversible extensibility. Superficial cuticle of some invertebrates, e.g. earthworm, is a protein like collagen. Leather is tanned, i.e. fixed, collagen of dermis. See *Fibroblast*, *Reticulin Fibres.*

COLLAR-CELL. Choanocyte (q.v.).

COLLATERAL BUNDLE. See *Vascular Bundle.*

COLLEMBOLA. Springtails; order of small wingless insects (see *Apterygota*) which jump by sharply extending the abdominal appendages.

COLLENCHYMA. Tissue providing mechanical support to young, actively growing, plant structures. Consisting of living cells with walls strengthened by cellulose thickening, usually in the corners; still capable of extension. Commonly found in cortex of herbaceous stems.

COLON. Large intestine of vertebrates, excluding the narrower terminal rectum. In amniotes and some Amphibia, but not in fish, clearly marked off from small intestine by valve, and probably largely concerned with absorption of water from faeces.

COLONIAL ANIMAL. Kind of animal which is organized as associations (colonies) of incompletely separate individuals. E.g. many Hydrozoa, Polyzoa.

COLOSTRUM. Maternal milk of mammal formed during the first few days after the birth. Particularly rich in proteins, including antibodies in some mammals, especially ungulates.

COLUMELLA. (1) Dome-shaped structure present in sporangia of many phycomycete fungi of the order Mucorales (pin moulds), produced by formation of convex septum cutting off sporangium from hypha bearing it. (2) Sterile central tissue of moss capsule.

COLUMELLA AURIS. Rod of bone or cartilage connecting ear-drum' to inner ear, and transmitting sound, in reptiles, birds, and Anura. homologous with hyomandibula of fishes, stapes of mammals.

COMMENSAL. (Of members of different species) living in close association, e.g. in same burrow, shell, or house, without much mutual influence, i.e. not symbiotic.

COMMISSURE. (1) Transverse cord of nerve-fibres uniting members of each pair of ganglia in double nerve-cord of Arthropoda and Annelida. (2) *Circum-oesophageal commissure*, nerve cord connecting ganglia in head above oesophagus with those below it, in e.g. Arthropoda and Annelida. (3) Bundle of nerve-fibres connecting right and left sides of brain or spinal cord in vertebrates. (4) Nervous cords uniting ganglia of Molluscs.

COMMON. (Of a vascular bundle), passing through stem and leaf. Cf. *Cauline*.

COMMUNITY. Ecological term for any naturally occurring group of different organisms inhabiting a common environment, interacting with each other especially through food relationships, and relatively independent of other groups. Communities may be of varying sizes, and larger ones may contain smaller ones. See *Association, Consociation, Society*.

COMPANION CELLS. Small cells, characterized by dense cytoplasm and prominent nucleus, lying side by side with sieve-tube cells in phloem of flowering plants and arising with them by longitudinal division of common parent cells.

COMPENSATORY HYPERTROPHY. Increase in size of residual part of a tissue or organ, some of which has been removed or put out of action, e.g. when one kidney is removed other enlarges. See *Hypertrophy, Regeneration*.

COMPETENCE. Of embryonic tissues; ability to react to a stimulus which causes development in a particular direction. Before determination (q.v.) tissues are usually competent to develop in a number of different directions. When tissues become determined, they lose their competence.

COMPETITIVE INHIBITION. Where two processes compete for some raw material which both use, inhibition of one process by diversion of available supplies of the raw material to the other. Particularly used of competing enzyme systems.

COMPLEMENT. Substance (a protein) normally present in blood which takes essential part in the cytolysis produced by certain antibodies in bacteria or other cells. There is only one kind of complement for all these reactions. Complement disappears from serum during most antigen-antibody reactions: this 'complement-fixation' can be used to test for presence of an antigen or an antibody, and is basis of Wasserman test for syphilis.

COMPLEMENTAL MALES. Males which live attached to females, and are usually small and more or less degenerate, except for reproductive organs; found in e.g. certain Crustacea, angler-fish.

CONCENTRIC BUNDLE. See *Vascular Bundle.*

CONCEPTACLE. Cavity containing sex organs, occurring in groups on terminal parts of branches of thallus in some brown Algae, e.g. bladder wrack.

CONDITIONED REFLEX. A reflex (q.v.) modified by experience, the original sensory component (i.e. the stimulus, sense organ, and sensory nerve path involved in it) being replaced by a different sensory component, the motor component (i.e. the response) remaining unchanged. E.g. food placed in a dog's mouth evokes reflex flow of saliva. If the introduction of food is accompanied by ringing a bell, and this is repeated several times, the bell alone eventually becomes able to evoke the salivation; though the effect is temporary unless reinforced from time to time by administering both food and bell. The original salivation reflex, which is inborn, i.e. does not depend on the experience of the animal, is the *unconditioned reflex*; it is not lost as a result of the experiment. The induced response to the bell alone is a conditioned reflex.

CONDUCTION. Passage of a physiological disturbance through a cell or tissue as a result of stimulation at one point, e.g. conduction of nerve impulse (q.v.).

CONDYLE. Ellipsoid knob of bone, which fits into a corresponding socket of another bone, the condyle and its socket forming a joint allowing movement in one or two planes but no rotation; e.g. condyle at each side of lower jaw, where it articulates with skull; occipital condyles on skull of tetrapods, fitting into atlas vertebra.

CONE. (Bot.). (*Strobilus*). Reproductive structure consisting of a number of sporophylls (q.v.) more or less compactly grouped on a central axis, e.g. cone of pine tree. (Zool.). Kind of light-sensitive nerve-cell present in retina of most vertebrates (though usually not in those which live in darkness). There are about 6 million in the retina of a primate. Concerned in colour discrimination, and in the most acute discrimination of detail. See *Fovea, Rod.*

CONIDIOPHORE. See *Conidium.*

CONIDIUM. Asexual spore of certain fungi cut off externally at apex of specialized hypha (conidiophore).

CONIFERALES. Order of Gymnospermae including extinct forms and nearly all living gymnosperms, pine, spruce, cedar, yew, larch, etc.; characteristic of temperate regions, majority tall, evergreen, forest trees. Usually monoecious with distinct male and female cones. Microsporophylls borne directly on cone axis; female cones compound, ovules borne on ovuliferous scales in the axils of bracts arising from the cone axis.

CONJUGATION. (Bot.) Type of sexual reproduction in which contents of two morphologically similar cells behave as isogamous gametes, neither being free-swimming, e.g. in green alga *Spirogyra*. (Zool.). (1) Union of gametes, free-swimming or not, especially in isogamy (q.v.). (2) Process by which sexual reproduction is effected in most Ciliata. In the simplest cases two individuals partially fuse; macronuclei disintegrate and micronuclei undergo changes, including meiosis, resulting in production of two gamete nuclei from each; one gamete nucleus from each organism passes into the other organism and there fuses with the stationary gamete nucleus, so that a zygote nucleus results in each individual. The organisms then separate, the zygote nucleus undergoes further divisions and ultimately gives rise to a micronucleus and macronucleus. (3) Also used more loosely as synonymous with syngamy.

CONJUNCTIVA. The layer of mucus-secreting epidermis, with underlying connective tissue, covering the white of the eye and lining the eyelids in vertebrates.

CONNECTIVE TISSUE. Vertebrate tissue consisting of *connective tissue fibres* usually mainly of collagen (q.v.), but with varying amounts of elastin and reticulin; scattered cells (fibroblasts and macrophages); blood and lymph vessels; tissue fluid in spaces; and an amorphous polysaccharide-containing matrix. Supporting or packing in function. Many modifications occur. See *Areolar Tissue*, *Adipose Tissue*, *Fascia*. See Fig. 2.

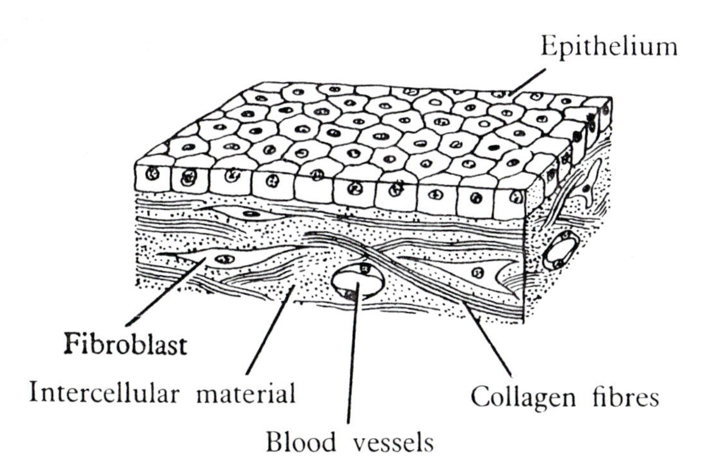

Fig. 2. Diagram of a piece of connective tissue covered with epithelium

CONSOCIATION. (Of plants), unit of natural vegetation dominated by *one* particular species, e.g. beechwood, dominated by the common beech tree. Cf. *Association*.

CONTACT INSECTICIDE. Poison which kills insects without their eating it, by e.g. penetrating body surface or blocking spiracles.

CONTOUR FEATHERS. Feathers with barbules (q.v.), forming smooth surface of bird.

CONTRACTICLE VACUOLE. Of Protista, particularly in many fresh water species, and of fresh water sponges, vacuole which periodically expands, filling with water, and then suddenly contracts, expelling its contents to exterior of cell. It probably removes the water which continually enters the cell from the environment by osmosis.

CONVERGENCE. Evolution such as to produce an increasing similarity in some characteristic(s) between groups or organisms which were initially different.

COPEPODA. Subclass of Crustacea; some minute marine species occur in the plankton in such numbers as to be important food for fishes.

CORACOID. Bone of shoulder-girdle of vertebrates, on ventral side. Meets scapula at glenoid cavity. A cartilage-bone. Reduced to small process of scapula in mammals (except monotremes). See Fig. 10, p. 216.

CORAL. Calcareous skeleton or fused skeletons of various Coelenterates and Polyzoa or the skeleton together with the animals (polyps) which secrete it. The great bulk of most coral reefs is made by the madreporarians (Actinozoa).

CORD. See *Spinal Cord, Umbilical Cord.*

CORDAITALES. Order of extinct, palaeozoic Gymnospermae (q.v.) that flourished particularly during the Carboniferous (q.v.). Tall, slender trees with dense crown of branches bearing many large, simple, elongated leaves. Sporophylls distinct from vegetative leaves, much reduced in size and arranged compactly in small, distinct, male and female cones. Microsporophylls stamen-like, interspersed among sterile scales; megasporophylls similarly borne, each consisting of a stalk bearing a terminal ovule.

CORIUM. Dermis (q.v.).

CORK (PHELLEM). Protective tissue of dead, impermeable cells formed by cork cambium (phellogen), which, with increase in diameter of young stems and roots, replaces the epidermis. Developed abundantly on the trunk of certain trees, e.g. cork oak, from which it is periodically stripped for commercial use.

CORM. (Bot.). Organ of vegetative reproduction; swollen stem base containing food material and bearing buds in the axils of scale-like remains of leaves of previous season's growth, e.g. crocus, gladiolus. Cf. *Bulb.*

CORNEA. (1) The transparent epidermis and connective tissue at front surface of the eye of vertebrates, overlying iris and lens. Mainly responsible in land vertebrates for the refraction which results in focusing an image on retina.

CORNIFICATION. Conversion of cells into keratin, e.g. in epidermis.

COROLLA. Usually conspicuous, often coloured part of a flower, within the calyx, consisting of a group of petals. See *Flower.*

CORONARY VESSELS. Arteries and veins of vertebrates carrying blood supply of heart muscle.

CORPORA ALLATA. Small glandular organs in the head of insects, known to produce hormones in some species.

CORPUS LUTEUM. Temporary organ of internal secretion, of the hormone progesterone (q.v.). Formed in mammals in interior of a ruptured Graafian follicle after ovulation, by ingrowth of follicle wall, which becomes yellow secretory *luteal tissue* (corpus luteum means 'yellow body'). Formation occurs as a result of action of luteinizing hormone (q.v.) of pituitary. If ovulation does not result in fertilization, then corpus luteum soon degenerates (see *Oestrous Cycle*). If fertilization occurs, corpus luteum persists and continues secreting during part or all of pregnancy. Bodies similar to corpus luteum occur in viviparous vertebrates other than mammals.

CORTEX. An outer layer or rind. (1) Of plants, parenchyma tissue surrounding the vascular cylinder (q.v.) in stems and roots and bounded on the outside by the epidermis (q.v.). (2) *Adrenal c.*, see *Adrenal Gland.* (3) Of *brain*, see *Cerebral Cortex.* (4) Of *cell*, see *Ectoplasm.*

CORTICOTROPIN. ACTH (q.v.).

CORTISONE. One of the hormones produced by the cortex of the adrenal gland (q.v.) but in unimportant amounts compared with other cortical hormones. 17-hydroxy–11-dehydrocorticosterone. Has the usual complex effects of cortical hormones, and notably produces a diminution of local inflammation and healing responses.

CORYMB. Kind of inflorescence. (q.v.).

COSMOID SCALE. Scale typical of primitive Choanichthyes. Three layers; outer like a layer of fused teeth (denticles), consisting of dentine with pulp cavities and thin superficial covering of modified, enamel-like dentine (ganoine); middle of loose, vascular bone; inner layer of bone in compact laminated sheets. Scale grows thicker by adding to inner layer only (cf. *Ganoid scale*).

COSTAL. Concerned with rib.

COTYLEDON. Leaf forming part of embryo (q.v.) of seeds; much simpler in structure than later formed leaves and usually lacking chlorophyll. Monocotyledons have one, dicotyledons two cotyledons in each seed. In gymnosperms the number varies. Cotyledons play an important part in the early stages of seedling development. In some seeds, e.g. peas, beans, they are storage organs from which the seedling draws food; in other seeds, e.g. grasses, food stored in another part of the seed, the endosperm (q.v.), is absorbed by the cotyledons and passed on to the seedling. In addition, the cotyledons of many plants later appear above ground, develop chlorophyll, and synthesize food material by photosynthesis (q.v.).

COVERSLIP. Extremely thin (about $\frac{1}{6}$ mm.) piece of glass used to cover specimens (sections, whole mounts, etc.), which have been prepared in Canada balsam (or similar substance) on a slide for microscopical examination.

COXA. Basal segment of insect leg.

CRANIAL NERVE. Peripheral nerve emerging from brain of vertebrate, i.e. originating within the skull; as distinct from spinal nerve, which emerges from spinal cord. Dorsal and ventral roots (see *Nerve Root*) of several segments exist amongst these nerves, but (unlike those of spinal cord) remain separate. Each root is numbered and named as a separate nerve, but the numbering bears little relation to segmentation of head. Olfactory and optic nerves are not equivalent to either dorsal or ventral roots. There are ten cranial nerves on each side in living anamniotes, eleven or twelve in amniotes and fossil amphibians. Include nerves supplying eyes, ears, nose, jaw muscles, skin of face, etc. See *Olfactory, Optic, Oculomotor, Trochlear, Trigeminal, Abducens, Facial, Auditory, Glossopharyngeal, Vagus, Accessory, Hypoglossal nerves.*

CRANIATA. Vertebrata (q.v.).

CRANIUM. Skull (of vertebrates). See *Neurocranium, Splanchnocranium.*

CRETACEOUS. Geological period (q.v.) lasting approximately from 135 till 70 million years ago.

CRINOIDEA. Feather stars, sea-lilies. Class of Echinodermata. Have long, branched, feathery arms. Usually sedentary and stalked, but the only British form (*Antedon*) is free as adult, sedentary as larva. Long and important fossil history, from Cambrian onwards.

CRITICAL GROUP. Group of related organisms in which, though variations exist, it is difficult to detect and define units suitable for taxonomic recognition.

CROCODILIA. Alligators and crocodiles. An order of Reptilia, with internal openings of the nose far back in the mouth owing to the presence of a long bony false palate (q.v.) (as in mammals).

CROP. In vertebrates, expanded part of oesophagus in which food is stored, e.g. in pigeon; in invertebrates, an expanded part of gut near head end in which food may be stored or digested.

CROP MILK (PIGEON'S MILK). Secretion of crop epithelium of male and female pigeon on which nestlings are fed. Like mammalian milk, production is strongly influenced by hormone prolactin (q.v.).

CROSS. Act or product of cross-fertilization.

CROSS-FERTILIZATION. Fusion of male and female gametes from different individuals of the species. Cf. *Self-Fertilization.*

CROSSING-OVER. Failure of linkage (q.v.) between genes of the same linkage-group. Due to chiasmata, i.e. mutual exchange of parts containing corresponding loci, between homologous chromosomes during their pairing at meiosis (q.v.). See Fig. 6b, p. 148.

CROSSOPTERYGII. Sub-class of Choanichthyes. Fish, almost all freshwater. Many fossil forms, which include ancestors of land vertebrates, but only one known living genus (*Latimeria* q.v.). Differ from Dipnoi in having normal conical teeth. First appeared in Devonian (about 400 million years ago), and their known fossil record ceases 70 million years ago.

CROSS-OVER VALUE (C.O.V.). Frequency of crossing over between two genes in different parts of the same chromosome, expressed as percentage of gametes in which one of the two genes has been exchanged for an allele from the homologous chromosome.

CROSS-POLLINATION. Transference of pollen from stamens of a flower to stigma of a flower of another plant of the same species. Cf. *Self-pollination.*

CRUSTACEA. Class of Arthropoda, including shrimps, crabs, water-fleas, etc. Mostly aquatic; two pairs of antennae; one pair of mandibles; other limbs biramous. See *Cirripedia, Cladocera, Copepoda, Amphipoda, Isopoda, Decapoda.*

CRYOPHYTES. Plants growing on ice and snow; micro-plants, largely consisting of Algae but including also some mosses, fungi, and bacteria. Algal forms may be so abundant as to colour substratum, e.g. 'red snow' due to presence of species of *Chlamydomonas.*

CRYPTIC COLORATION. Coloration of animals which conceals by resemblance to surroundings.

CRYPTOGAMIA. Name given by early systematic botanists to a group of plants consisting of Thallophyta, Bryophyta, and Pteridophyta; so called because the organs of reproduction in these plants are not prominent as they are in flowering plants (Phanerogamia).

CRYPTOZOIC (Of animals). Inhabiting crevices, e.g. under stones, leaves.

C.S.F. Cerebrospinal fluid (q.v.).

CTENOPHORA. Comb-jellies, sea-gooseberries, Venus' girdle, etc. Subphylum of Coelenterata (sometimes considered distinct phylum). Without thread cells; body neither polyp nor medusa; move by cilia fused in rows forming 'combs' or 'ctenes'; no vegetative reproduction; no sedentary phase. Cf. *Cnidaria.*

CUSP. Projection on biting-surface of mammalian molar tooth.

CUTICLE. Superficial non-cellular layer covering animal or plant; secreted by epidermis. In *higher plants* forms a continuous layer over aerial parts broken only by stomata and lenticels; protects against mechanical injury, but chief function is preventing excessive water-loss. See *Cutin.* In *Metazoa* it is present in most invertebrates, mainly made of collagen-like protein, or of chitin. In Arthropoda it is firm enough to act as skeleton and is composed of chitin and protein, hardened by lime-salts in Crustacea. In insects, it has a very thin, fairly waterproof, covering which prevents excessive water-loss. In vertebrates, stratum corneum (q.v.), which is cellular, is occasionally called cuticle.

CUTICULARIZATION. Process of cuticle formation.

CUTIN. Complex mixture of oxidation and condensation products of fatty acids of which plant cuticle (q.v.) is composed.

CUTINIZATION. Impregnation of plant cell wall with cutin.

CUTIS. Skin; in vertebrates composed of epidermis and dermis.

CUTTING. Artificially detached part of plant used as a means of vegetative propagation.

CUVIERIAN DUCT. Ductus Cuvieri (q.v.).

CYANOPHYCEAE (MYXOPHYCEAE). Blue-green algae. Small class of Algae characterized by blue-green colour due to presence of pigment phycocyanin distributed with chlorophyll throughout cytoplasm and masking its colour. Widely distributed, aquatic or in damp places, abundant in freshwater plankton and often in soil. Primitive algae with simple cell structure, lacking a true nucleus. Microscopic, unicellular, occurring singly or grouped into colonies. Reproduction entirely asexual, by cell division or fragmentation of a colony.

CYCADALES. Order of Gymnospermae with fossil (Mesozoic), and a few living, representatives. Most primitive living seed plants. Indigenous to tropical and sub-tropical zones; living to a great age, up to 1000 years. Stem unbranched, tuberous or columnar, up to 20 metres in height, bearing a crown of fern-like leaves. Dioecious. Microsporophylls distinct from vegetative leaves, arranged in a compact cone. Megasporophylls leaf-like structures, loosely grouped, or highly modified, grouped in a compact cone. Male gametes motile by cilia.

CYCADOFILICALES (PTERIDOSPERMAE). Order of extinct, palaeozoic Gymnospermae (q.v.) that flourished particularly during the Carboniferous (q.v.). Of great phylogenetic interest. Reproducing by seeds but possessing fern-like leaves and internal anatomy combining fern-like vascular system with development of secondary wood. Micro- and megasporophylls little different from ordinary vegetative fronds, not arranged in cones.

CYCLOPIA. Having a single median eye; sometimes occurs in vertebrates through defective induction (q.v.) of eyes by head mesoderm during embryonic development.

CYCLOSIS. (Bot.). Circulation of protoplasm in cells.

CYCLOSTOMATA. (1) Order of class Agnatha. Consists of lampreys and hagfish, the only living representatives of Agnatha (see *Ostracodermi*). Eel-like aquatic animals, with jawless sucking mouth, single nostril, no bone or scales, and no paired fins. They attach themselves by their mouths to fish, whose flesh they rasp with horny tongue, sucking the blood. (2) Sub-order of Polyzoa.

CYME. Kind of inflorescence (q.v.).

CYPSELA. Characteristic fruit of the family Compositae (sunflower, daisy, etc.). Like an achene (and usually so described) but formed from an inferior ovary and thus sheathed with other floral tissues outside ovary wall. Strictly a pseudo-nut as it is formed from two carpels.

CYSTICERCOID. Larva of certain tapeworms, cysticercus-like but with small bladder.

CYSTICERCUS (BLADDER WORM). Larval form of some tapeworms,

consisting of a scolex tucked into a large bladder (cf. *Cysticercoid*, *Plerocecoid*); occurs in intermediate host; grows into mature tapeworm if swallowed by definitive host. See *Onchosphere*.

CYSTOCARP. Structure developed after fertilization in Red Algae; consisting of filaments bearing terminal carpospores (q.v.), produced from the fertilized carpogonium, the whole being enveloped in some genera by filaments arising from neighbouring cells.

CYSTOLITH. Stalked body consisting of ingrowth of cell wall bearing deposit of calcium carbonate; found in epidermal cells of certain plants, e.g. stinging nettle.

CYTOCHROME. Mixture of similar substances (proteins, with an iron-containing prosthetic group allied to that of haemoglobin) concerned, mostly as coenzymes, in cell respiration; widely distributed in aerobic organisms. Wing muscles of insects are particularly rich in it. Its oxidation by molecular oxygen, and subsequent reduction by oxidizable substances in the cell (both processes under influence of enzymes), is main way in which atmospheric oxygen enters into metabolism of cell; about 90 per cent of oxygen consumption is controlled by cytochrome. The cytochromes have characteristic absorption spectra, which makes it possible to study their reactions in living cells and tissues.

CYTOGENETICS. Science which integrates the methods and findings of cytology and genetics.

CYTOLOGY. Study of cells.

CYTOLYSIS. Dissolution of cells, particularly by destruction of their surface membranes.

CYTOPLASM. All the protoplasm of a cell excluding the nucleus. It is usually a transparent slightly viscous fluid with inclusions of various sizes ranging from microscopically visible plastids and Golgi apparatus to invisible ribosomes. Differentiated externally as a plasma-membrane (q.v.). See Fig. 1, p. 44.

CYTOPLASMIC INHERITANCE. Non-mendelian inheritance by self-propagating cytoplasmic entities; as distinct from gene (q.v.) inheritance. See *Plasmagene*, *Plastogene*.

CYTOTAXONOMY. (Bot.). Classification based on characters of somatic chromosomes (number, size, and shape).

D

DARWINISM. Theory of evolution by natural selection (q.v.).

DAUGHTER CELLS (D.-NUCLEI). The two cells (or nuclei) resulting from division of a single cell, usually by mitosis.

DEAMINATION. Removal of amino (NH_2) group. In mammals occurs to many amino-acid molecules by action of deaminating enzymes in liver and kidney; the liberated ammonia being turned into urea by the liver, the carbon-containing residue probably providing energy by being oxidized.

DECAPODA. (1) Order of Arthropoda, class Crustacea, including crabs, lobsters, crayfish, shrimps. (2) Sub-order of Mollusca, class Cephalopoda; squids, cuttlefish, etc.

DECIDUA. Mucous membrane (endometrium) lining uterus in the thickened and modified form it acquires during pregnancy in many mammals (not in 'ungulates', see *Placenta*). Some or all of the decidua comes away with the placenta at birth.

DECIDUOUS. (Of plants) shedding leaves at a certain season, e.g. autumn. Cf. *Evergreen.*

DECIDUOUS TEETH. Milk teeth. First of the two sets of teeth which most mammals have; similar to the second (permanent) set which replaces it, except in having grinding teeth corresponding only to the premolars, not to the molars, of that set.

DEFICIENCY DISEASE. Disease due to deficiency of some essential food substance, e.g. vitamin, mineral element, or essential amino acid.

DEGENERATION. Of an organ, loss of whole or part during life-cycle or in course of evolution. Of cells, death with accompanying changes. Of nerves, death and destruction of nerve fibres but not of their supporting cells. Evolutionary, loss of structure and function leading to a vestigial organ (q.v.).

DEGLUTITION. Swallowing, a complex reflex set off by stimulation of pharynx, involving contraction of muscles of mouth and pharynx, closing of the back of the nose by soft palate, raising of larynx against epiglottis, inhibition of breathing, peristalsis in oesophagus.

DEHISCENT. (Of fruits) opening to liberate the seeds, e.g. pea, viola, poppy.

DEHYDRATION. (In preparation of tissues for microscopical examination) the elimination of water, usually by soaking in successively stronger concentrations of ethyl alcohol. An essential preliminary to embedding (q.v.) in wax, or placing in Canada balsam (q.v.), both of these substances being quite immiscible with water. See *Clearing*, *Microtome.*

DEHYDROGENASE. Enzyme which catalyses oxidation of substrate by removing hydrogen from it. The hydrogen may combine with molecular oxygen (see *Oxidase*) or with another substance. Most oxidizing enzymes are dehydrogenases.

DEME. Basic unit in terminology of categories in experimental taxonomy (q.v.). Any specified population of plants possessing clearly definable genetical, cytological, or other characteristics, e.g. *gamodeme*, group of individuals which are capable of interbreeding.

DENATURATION Structural change produced in soluble (globular) proteins by mild heat or various chemicals, rendering them less soluble than in their original ('native') condition, involving an unfolding of the peptide chains.

DENDRITES. Branching cytoplasmic projections of a nerve cell (q.v.) which have synapses with, and receive impulses from, axons of other nerve cells.

DENITRIFYING BACTERIA. Soil bacteria which in absence of oxygen break down nitrates and nitrites with evolution of free nitrogen. See *Nitrogen Cycle.*

DENTAL FORMULA. Formula indicating, for a given species of mammal, the number of each kind of its teeth. The number in the upper jaw of one side is written above that in the lower jaw of one side; and the categories are given in the order: incisors, canines, premolars, molars. The formula of the primitive mammal is $i\frac{3}{3}$. $c\frac{1}{1}$. $p\frac{4}{4}$. $m\frac{3}{3}$. The human formula is $i\frac{2}{2}$. $c\frac{1}{1}$. $p\frac{2}{2}$. $m\frac{3}{3}$.

DENTARY. The only bone of the lower jaw of mammals, one on each side; one of several of the lower jaw bones in other vertebrates. A dermal bone.

DENTICLES (PLACOID SCALES). Tooth-like scales present in many fish. They completely cover body of elasmobranchs. Composed of dentine, with pulp-cavity. Teeth are probably modified denticles.

DENTINE. Main constituent of teeth. Like bone in structure, but contains no cells, though cell-processes penetrate it from cells in pulp cavity of tooth. Formed from mesoderm or neural crest of embryo. Ivory is dentine.

DEOXYRIBOSENUCLEIC (DESOXYRIBOSENUCLEIC) ACID (DNA). See *Nucleic Acid.*

DEPENDENT DIFFERENTIATION. Of an embryonic tissue, differentiation which is dependent on a stimulus coming from some other tissue. See *Competence, Induction, Organizer, Self-Differentiation.*

DERMAL BONE (MEMBRANE BONE). Bone arising near surface of embryo (though it may later sink deeper) from concentration of mesenchyme cells, with no cartilage precursor (see *Cartilage Bone*). Occurs in skull and shoulder-girdle of most vertebrates, e.g. roofing bones of skull; dentary; clavicle. In some very early vertebrates it formed extensive superficial armour, e.g. Ostracodermi.

DERMAPTERA. Small order of exopterygote insects, including earwigs, with fan-like hind wings folding under short stiff forewings.

DERMATOGEN. See *Apical Meristem.*

DERMIS (CORIUM). Innermost of the two layers of the skin of vertebrates, the outer being the epidermis (q.v.). Much thicker than epidermis. Consists of connective tissue, with abundant collagen fibres mainly parallel to surface; scattered cells; blood and lymph vessels; sensory nerves. Sweat-glands and hair-follicles project down from epidermis into dermis. Responsible for tensile strength of skin. May contain scales or bone.

DERMOPTERA. 'Flying lemurs', though it would be more appropriate to call them gliding insectivores. Small order of mammals with hairy membrane stretching along side of body, between wrist, ankle, and tip of tail.

DETERMINED. Term applied to embryonic tissues, meaning that in all conditions in which they can develop at all, they will form only one particular sort of adult tissue or organ, i.e. their fate is fixed. See *Competence*, *Presumptive*.

DETRITUS. Organic debris from decomposing plants and animals.

DEUTOPLASM. Yolk.

DEVONIAN. Geological period (q.v.); lasted approximately from 400 till 350 million years ago.

DEXTRIN. Polysaccharide carbohydrate formed as an intermediate product in hydrolysis of starch to glucose.

DEXTROSE. Glucose (q.v.).

DIADELPHOUS. (Of stamens) united by their filaments to form two groups, or having one solitary and the others united, e.g. pea. Cf. *Monadelphous*, *Polyadelphous*.

DIAGEOTROPISM. Orientation of plant part by growth curvature in response to stimulus of gravity so that its axis is at right-angles to direction of gravitational force, i.e. horizontal; exhibited by rhizomes of many plants.

DIALYSIS. Method of separating small molecules, e.g. salts, from colloids, e.g. proteins, in mixed solution, by putting mixture in bag made of a membrane, e.g. collodion, permeable to small but not to large molecules, with excess water outside the bag into which small molecules diffuse.

DIAPAUSE. Of some species of insects, period of suspended development or growth, accompanied by greatly decreased metabolism. Often correlated with seasons, e.g. it may be hibernation (q.v.).

DIAPHRAGM. Sheet of tissue, part muscle, part tendon, covered by serous membrane, separating cavities of thorax (occupied by lungs and heart) from cavity of abdomen (occupied by intestine, liver, etc.). Present only in mammals. Diaphragm is arched up into thorax at rest and its flattening is a very important part of mechanism of inspiration of breath in most mammals.

DIAPHYSIS. Shaft of a long limb-bone, or central portion of a vertebra, in mammals. See *Epiphysis*.

DIASTASE. Amylase (q.v.).

DIASTOLE. Phase of heart-beat when heart muscle relaxes, and heart refills with blood from veins; or of contractile vacuole (q.v.) when it refills with fluid. Cf. *Systole*.

DIATOMS. Bacillariophyceae (q.v.).

DICHASIUM. Kind of inflorescence (q.v.).

DICHLAMYDEOUS (DIPLOCHLAMYDEOUS). (Of flowers), having perianth segments in two whorls.

DICHOGAMY. Condition in which male and female parts of a flower mature at different times, ensuring that self-pollination does not occur. See *Protandrous*, *Protogynous*. Cf. *Homogamy*.

DICHOTOMOUS. Dividing into two equal branches.

DICOTYLEDONEAE. Larger of the two classes into which flowering plants (Angiospermae) are divided; distinguished from the smaller class, *Monocotyledoneae*, by presence of two seed leaves (cotyledons) in the embryo and by other structural features, e.g. net-veined leaves; stem vascular tissue in form of a ring of open bundles; flower parts in fours or fives or in multiples of these numbers. Includes many forest trees, oak, elm, beech, etc.; fruit trees, food plants, e.g. potatoes, beans, cabbage; and ornamentals, e.g. rose, clematis, snapdragon.

DICTYOSTELE. Siphonostele (q.v.) that is broken up by crowded leaf gaps into a network of distinct vascular strands (meristeles), each surrounded by an endodermis. Present in stems of certain ferns.

DIDELPHIA. Marsupialia (q.v.).

DIDYNAMOUS. (Of stamens), four in number in two pairs, those of one pair longer than those of the other, e.g. foxglove.

DIFFERENTIATION. Change in structure and/or function of cells, tissues, or organs during development (embryonic or regenerative); usually resulting in an initially simple structure and/or function becoming transformed into one of the cell, tissue, or organ types of the normal adult.

DIGESTION. Breakdown of complex foodstuffs by enzymes to simpler compounds which can be incorporated into metabolism. In many animals it is *extracellular*, i.e. enzymes are secreted into gut cavity, in which digestion takes place, e.g. vertebrates, arthropods; in others it is wholly or partly *intracellular*, the cell of the gut phagocytosing solid food particles, e.g. coelenterates, lamellibranchs.

DIGIT. Finger or toe of pentadactyl limb (q.v.). Contains phalanges. See Fig. 7, p. 175.

DIGITIGRADE. Walking on toes, not on whole foot; i.e. on ventral surface of digits only, e.g. cat, dog. Cf. *Plantigrade*, *Unguligrade*.

DIKARYON (of fungus hypha or mycelium) composed of cells containing two haploid nuclei which undergo simultaneous division during formation of each new cell; characteristic of phase (dikaryophase) of life-cycle of many Basidiomycete fungi. Cf. *Monokaryon*.

DIKARYOPHASE. See *Dikaryon*.

DIMORPHISM. Existing in two forms, e.g. submerged and aerial leaves of water crowfoot; male and female individuals in animals.

DINOSAUR. Fossil reptile belonging to one of the two orders Saurischia and Ornithischia, into which the old order Dinosauria is now split.

DIOECIOUS. Unisexual, the male and female sexual reproductive organs borne on different individuals. Cf. *Monoecious*, *Hermaphrodite*.

DIPLOBLASTIC. Having the body made of two cellular layers only (ectoderm and endoderm). The Coelenterata are diploblastic, and all other Metazoa are triploblastic. In small simple Coelenterata such as *Hydra* the diploblastic condition is clear, but it is obscured (see *Mesogloea*) in large jelly-fish and sea-anemones. In diploblastic animals, however, unlike triploblastic, no *organs* develop from the middle layer, but only from the ectoderm and endoderm.

DIPLOID. (Of a nucleus) having the chromosomes in pairs, the members of which are homologous (q.v.), so that twice the haploid number is present. Characteristic of almost all animal cells except the gametes; of the zygotes of many green Algae and of many Fungi, which undergo meiosis so that later stages are haploid; and of the sporophyte of other Algae, Bryophyta and vascular plants. See *Alternation of Generations*, *Meiosis*, *Dikaryon*.

DIPLONT. Diploid stage of an organism, ending with meiosis. Cf. *Haplont*.

DIPLOPODA. Millipedes. Order or class of Arthropoda; see *Myriapoda*.

DIPNOI (DIPNEUSTI). Lung-fish. A sub-class of Choanichthyes. Fresh-water. Fairly common fossils, but only three living genera, all tropical: *Protopterus* (African), *Lepidosiren* (South American), *Epiceratodus* (Australian). Differ from Crossopterygii in their peculiar tooth plates, used to crush shells of small molluscs on which they feed. Have functional lungs, like Polypterus and many early fossil fish of other groups. First appear in Devonian (about 400 million years ago).

DIPTERA. Flies. Large order of endopterygote insects. Only one pair of wings, hind pair reduced to pegs (halteres, q.v.). Larva caterpillar-like or grub-like. Adults have mouth parts highly specialized, e.g. for sucking nectar of flowers, or blood. Some blood-suckers are of great medical and economic importance because they transmit certain diseases, e.g. mosquitoes, tse-tse flies.

DISACCHARIDE. A sugar, a compound of two monosaccharides (q.v.). The biologically important disaccharides have 12 carbon atoms (formed from two hexoses), e.g. sucrose, maltose, lactose.

DISTANCE RECEPTOR. Exteroceptor (q.v.) responding to stimuli (e.g. light, sound) which allows orientation of the animal to source of the stimuli.

DISTAL. Situated away from; especially from place of attachment. In a limb, away from the body; in a blood-vessel, away from the heart; in a nerve, away from the central nervous system. Cf. *Proximal*.

DIURESIS. Increased output of urine by kidney. E.g. occurs after drinking much water.

DIVERTICULUM. (Zool.). Blind-ending tubular or sac-like outpushing from a cavity.

DIVISION. Major group used in classifying plants. See *Phylum*.

DIZYGOTIC TWINS (FRATERNAL TWINS). Twins developed as

result of simultaneous fertilization of two separate ova. Such twins are genetically no more alike than siblings (q.v.). Cf. *Monozygotic twins.*

DNA. Desoxyribosenucleic acid. See *Nucleic Acid.*

DOMINANT. (1) (Genetics). A gene which produces the same character when it is present in single dose along with a specified allele (heterozygous), as it does in double dose (homozygous), is said to be dominant to that allele. The allele which is ineffective in the heterozygote is said to be *recessive* to that dominant. When a gene is called dominant without qualification it is usually implied that it is dominant to the normally occurring ('wild-type') allele(s). A dominance-recessiveness relation is common between two genes (the gene most frequently present at a given locus in a wild population is usually dominant to its alleles), but it is not universal. (2) (Plant Ecology). Of a plant species, the most characteristic species in a particular plant community, to a large extent governing the type and abundance of other species in the community, e.g. beech trees in a beechwood. When more than one dominant species occurs in a particular plant community, they are called *Co-dominants.*

DONOR. An individual from whom tissue is taken to transfer to another, e.g. in grafting; blood transfusion.

DORMANT. In a resting condition; alive, but with relatively inactive metabolism. Dormancy may involve the whole organism as in higher plants and animals, or be confined to reproductive bodies, e.g. resting spores, sclerotia of Fungi, spores of Bacteria, and in animals, resting eggs, statoblasts. Dormancy involves cessation of growth. It may be imposed by unfavourable conditions, e.g. low temperature, lack of moisture, ending as soon as the limiting conditions are removed; or be spontaneous, independent of external conditions. In higher plants, imposed dormancy occurs in many seeds, e.g. pea, wheat, which, though capable of germination immediately after harvesting, are dormant in the seedsman's hands. On the other hand seeds of other plants, e.g. hawthorn, *Daphne mezereum,* will not germinate immediately, even in conditions otherwise favourable to germination, but exhibit a period of spontaneous dormancy. Similarly deciduous trees and shrubs have a dormant period every winter. Although in general this coincides with unfavourable conditions, it is spontaneous. Such a rest period appears to be a part of an annual rhythm exhibited by higher plants. It is strikingly seen in bulbs, e.g. snowdrop, daffodil, where this spontaneous dormancy, following vegetative growth and flowering in spring, coincides with a period favourable to the growth of other plants. See also *Hibernation, Aestivation, Diapause.*

DORSAL. (Zool.). Situated at, or relatively nearer to, the back, i.e. the side of the animal which (if not in the animal, at least in the group to which the animal belongs) is normally directed upwards with reference to gravity. In human beings, it is directed backwards. Opposite of ventral. *D. aorta*, see *Aorta, Dorsal.* (Bot.). Also used of leaves, synonymous with Abaxial (q.v.).

DORSAL LIP. Of blastopore (q.v.). Part of rim of blastopore corresponding to future dorsal side, in Amphibian embryo. See *Organizer.*

DORSAL ROOT (POSTERIOR R., SENSORY R.). Nerve-root (q.v.) of vertebrates containing the sensory fibres.

DORSIVENTRAL LEAVES. Showing differences in structure between upper and lower sides. Characteristic of the more or less horizontal leaves of dicotyledons, e.g. sycamore. Below the upper epidermis, lying regularly side by side and with their long axes at right-angles to it, are one or more layers of elongated cells forming the palisade mesophyll. Between this and the lower epidermis is the spongy mesophyll, consisting of irregularly shaped cells with large intercellular spaces. Cf. *Isobilateral.*

DOUBLE FERTILIZATION. In Angiosperms union of one male nucleus with egg nucleus to form zygote and of the other with primary endosperm nucleus to form endosperm nucleus. See *Embryo sac.*

DOUBLE RECESSIVE. Individual which is homozygous for a particular recessive gene. To test whether an individual showing character of a dominant gene is homozygous or heterozygous, it is crossed to corresponding double recessive (constituting a *test-cross*); if homozygous, all offspring show dominant phenotype; if heterozygous, half show dominant and half recessive phenotypes.

DOWN-FEATHER. See *Barbules.*

DROSOPHILA. Fruit-fly or banana-fly. Genus of the order Diptera. Experiments on *Drosophila melanogaster* by T. H. Morgan and his school established the foundations of modern genetics in relation to chromosomes; and it has been widely used in genetical research ever since. See *Salivary Gland Chromosome.*

DRUPE. Succulent fruit in which wall consists of outer skin (*epicarp*), thick fleshy *mesocarp* and hard, stony *endocarp* enclosing a single seed. Commonly called stone-fruit, e.g. plum, cherry, Cf. *Berry.* In some plants, e.g. coconut, mesocarp is very fibrous. Pericarp here consists of tough, leathery epicarp, very thick fibrous mesocarp and hard endocarp enclosing seed and forming with it the 'nut' we buy.

DUCTLESS GLAND. Endocrine gland (q.v.).

DUCTUS CUVIERI (CUVIERIAN DUCT). Main vein of fish (and of tetrapod embryo) returning blood from cardinal veins (in body-wall) to heart (under gut) in a fold of coelomic lining which forms the posterior wall of pericardial cavity. There is one on each side. Becomes vena cava superior (q.v.) in adult tetrapods.

DUODENUM. First part of small intestine of vertebrates. Leads out of stomach (junction being guarded by pyloric sphincter). Receives bile and pancreatic ducts. A region of active digestion.

DURA MATER. Firm connective tissue covering brain and spinal cord of vertebrates, containing blood-vessels. See *Pia-arachnoid.*

E

EAR, INNER (MEMBRANOUS LABYRINTH). The sense-organ of vertebrates which detects position with respect to gravity; acceleration; and sound. Lies inside skull (in auditory capsule) connected by auditory nerve to brain. Consists of two connected sacs containing otoliths (q.v.). From one (utricle) arise semicircular canals (q.v.). From other (saccule) arises organ of hearing (see *Cochlea*). Whole inner ear is filled with a fluid (endolymph).

EAR, MIDDLE (TYMPANIC CAVITY). Cavity between ear-drum and auditory capsule, present in tetrapod vertebrates (not Urodela, Apoda, Snakes). Derived from a gill pouch (spiracle). Communicates with mouth by eustachian tube and is filled with air. Crossed by ear ossicles. Encased in bony projection of skull (bulla) in most mammals.

EAR OSSICLE. Bone in middle ear, connecting ear-drum to inner ear, in tetrapod vertebrates. Transmits vibrations of ear-drum caused by sounds to endolymph of inner ear, which in turn affects cochlea. A single bone, columella auris (q.v.), representing hyomandibula of fishes, is present in ear of reptiles, birds, and many amphibians. Three bones, malleus, incus, and stapes, of interesting evolutionary history, are present in mammalian ear. Stapes represents columella. Malleus represents a bone of lower jaw (articular) which in other vertebrates articulates with quadrate bone of upper jaw forming jaw-joint; but a new kind of joint in mammals (between dentary and squamosal) frees articular; and it also frees quadrate, which becomes incus. The three ossicles of mammals form a lever system, diminishing amplitude of sound waves and increasing force on inner ear.

EAR, OUTER OR EXTERNAL. Part of ear of tetrapod external to ear-drum. Absent in Amphibia and some reptiles, where ear-drum (if present) is at skin surface. In other tetrapods, ear-drum is at inner end of short tube (*external auditory meatus*); this, in mammals, with the flap of skin and cartilage (pinna) at outer opening of meatus, makes up outer ear.

ECAD. (Bot.). A habitat form, showing characteristics imposed by habitat conditions and non-heritable.

ECDYSIS. Moulting. In arthropoda, periodic shedding of cuticle (q.v.). In insects inner part of old cuticle is absorbed, the rest is split at line of weakness, and the insect draws itself out, clothed in a preformed soft new cuticle. By swallowing air the insect quickly increases its bulk and the new cuticle finally hardens a size larger than the old. The lining of all but the finest tracheae (q.v.) is shed with the old cuticle. Ecdysis is initiated by a hormone. It does not usually occur in adult insects. In reptiles, periodic shedding of outer layer of epidermis occurs in all, except crocodiles.

ECESIS. Germination and establishment of colonizing plants.

ECHIDNA. Spiny anteater, a very primitive monotreme mammal of New Guinea and Australia.

ECHINODERMATA. Phylum of animals containing sea-urchins, sea-cucumbers, starfish, brittle-stars, feather-stars, and sea-lilies. Marine; more or less radially symmetrical (with usually five axes of symmetry); calcareous skeletal plates in the skin; tube feet; coelom well developed, consisting of several intricate spaces, one of which is the water vascular system; larva (when present) more or less bilaterally symmetrical with coelom in three segments. Their mode of development shows them to be related to the same stock that gave rise to vertebrates. Includes classes Asteroidea, Ophiuroidea, Echinoidea, Holothuroidea, Crinoidea. Palaeozoic rocks contain remains of several other groups of echinoderms now extinct; they were mostly sedentary forms. It seems that modern free-living echinoderms were derived from radially symmetrical sedentary forms.

ECHINOIDEA. Sea-urchins, heart-urchins, cake-urchins; class of Echinodermata; more or less globular in shape; spiny; skeletal plates in skin joined to form firm skeleton.

ECOLOGY. Study of the relations of animals and plants, particularly of animal and plant communities, to their surroundings, animate and inanimate.

ECOSPECIES. Group of plants comprising one or more ecotypes (q.v.) within a coenospecies (q.v.) whose members can reproduce amongst themselves without loss of fertility in offspring. Approximates to conventional 'species'.

ECOTYPE. Group of plants within a species adapted genetically to a particular habitat but able to cross freely with other ecotypes of the same species.

ECTOBLAST (EPIBLAST). See *Ectoderm.*

ECTODERM (ECTOBLAST, EPIBLAST). Superficial germ-layer of animal embryo, developing mainly into epidermis, nervous tissue, and nephridia (when present). Term is usually applied to the germ-layer while it is still a distinctly demarcated region of the embryo, after gastrulation but before differentiation into the derived tissues; and all the tissues at later stages derived from the embryonic layer may be called ectodermal. See *Germ-Layer*, *Endoderm*, *Mesoderm.*

ECTOPARASITE. Organism living parasitically on the outside of another organism. E.g. flea. Cf. *Endoparasite.*

ECTOPLASM. (Bot.). (*Ectoplast*) Synonymous with external plasma membrane (q.v.). (Zool.). (*Cell cortex*). Outer layer of cytoplasm, which differs from inner cytoplasm (endoplasm) in many cells, though the two regions grade into each other. Unlike endoplasm, ectoplasm is usually semi-solid (a gel, in which case it may be called *plasmagel*) and often contains relatively few granules, e.g. in many ova and Protozoa. In planktonic Protozoa it may be highly vacuolated. Ectoplasm has important role in many forms of cell movement, including cell-division and amoeboid movement; but active as it is this is not the kind of ectoplasm which undertakes table-turning.

ECTOPROCTA. Sea-mats: group of small colonial aquatic animals sometimes regarded as a class of Polyzoa (q.v.), sometimes as a separate phylum.

ECTOTROPHIC. (Of mycorrhizas) with the mycelium of the fungus forming an external covering to the root, e.g. in pine trees. Cf. *Endotrophic*. See *Mycorrhiza*.

EDAPHIC FACTORS. Environmental conditions that are determined by the physical, chemical, and biological characteristics of the soil.

EDEMA. Oedema (q.v.).

EDENTATA. Three orders of placental mammals, with teeth reduced or absent, which used to be grouped together as one order. Sloths, armadilloes, American anteaters (Xenarthra); aard-vaark (Tubulidentata); scaly anteaters (Pholidota).

E.E.G. (ELECTROENCEPHALOGRAM). Record of the rhythmical waves of change in electrical potential occurring in vertebrate brain, mainly cerebral cortex, which can be detected through intact skull. Patterns in man can be correlated with various physiological and pathological states such as epilepsy.

EELWORMS. Nematodes; applied only to plant-parasitic and free-living forms.

EFFECTOR. Organ or cell by which an animal acts. Main effectors are muscles, glands, cilia, chromatophores, thread-cells.

EFFERENT. Leading away from. E.g. of arteries leading from vertebrate gills, or of arteriole from glomerulus; or of nerve fibres conducting impulses from central nervous system (motor fibres). Cf. *Afferent*.

EGG MEMBRANES. Protective membranes surrounding the eggs of animals, especially well-developed in those of terrestrial species. Classified according to mode of origin: (1) Those secreted by ovum itself, vitelline membrane, q.v.; (2) Those secreted by cells in the ovary, e.g. chorion, q.v., of insect eggs; (3) Those secreted by oviduct, e.g. white and shell of bird's egg, jelly of frog's spawn.

ELASMOBRANCHII. Often synonymous with Chondrichthyes (q.v.). Sometimes made a sub-class of Chondrichthyes, including all members of this class except Holocephali.

ELASTIN. A fibrous protein in the form of highly extensible and elastic fibres ('yellow fibres') found sparsely scattered in vertebrate connective tissue; numerous in some places, e.g. in lungs, walls of large arteries.

ELATER. (1) Elongated cell with wall reinforced internally by one or more spiral bands of thickening, numbers of which occur among spores in capsules of liverworts; assist in discharge of spores by their movements in response to changes in humidity. (2) Appendage of spores of horsetails; formed from outermost wall layer, coiling and uncoiling as air is dry or moist, possibly assisting in spore dispersal.

ELYTRON (PLUR. ELYTRA). Modified front wing of beetles (Coleoptera). Thick and tough, at rest it protects the thin hind-wing folded beneath it.

EMASCULATION. Removal of stamens from hermaphrodite flowers before they have liberated pollen, as a preliminary to artificial hybridization.

EMBEDDING. See *Microtome.*

EMBRYO. (Bot.). Young plant developed after sexual or parthenogenetic reproduction from an ovum. In seed plants the embryo is contained in the seed and consists of a stem bearing a bud (plumule), a root (radicle), and one or more seed leaves (cotyledons). (Zool.). Animal in process of development from fertilized (or parthenogenetically activated) ovum. The embryo is contained in egg-membranes (q.v.), or, in viviparous species, in maternal body; and termination of embryonic period is usually taken to be hatching from membranes, or birth.

EMBRYO SAC. Large oval cell in nucellus of ovule of flowering plants in which fertilization of egg and development of embryo occurs; female gametophyte containing a number of nuclei derived by division of megaspore (q.v.) nucleus. At micropylar end, the *egg-apparatus*, consisting of egg nucleus and two others, *synergidae*; at opposite end three nuclei which become separated by cell walls forming *antipodal* cells, probably aiding in nourishment of young embryo; in centre, two *polar* nuclei which fuse to form *primary endosperm nucleus*. At fertilization one male nucleus fuses with egg nucleus forming a zygote which develops into embryo; second male nucleus fuses with primary endosperm nucleus to form *endosperm* nucleus, which divides to form endosperm.

EMBRYOLOGY. Study of development of embryos.

EMBRYONIC MEMBRANE. Extraembryonic membrane (q.v.).

ENAMEL. Hard covering of exposed part (crown) of teeth. Occurs also on denticles. Formed by epithelium of mouth, unlike dentine (q.v.). Consists mainly of crystals of a calcium phosphate-carbonate salt, with very little (3-5 per cent) organic substance.

ENATION. Outgrowth produced by local hyperplasia (q.v.) on leaf as a result of virus infection.

ENCYSTED. Surrounded by a cyst or shell.

ENDEMIC. Confined to a given region, e.g. an island or country. (Of pest or disease-producing parasites), continuously occurring in a particular area.

ENDOBLAST (ENTOBLAST, HYPOBLAST). See *Endoderm*.

ENDOCRINE GLAND. (Zool.). Gland producing hormones (q.v.).

ENDOCRINOLOGY. Study of hormones, their production, nature, and effects.

ENDODERM (ENTODERM, ENDOBLAST, ENTOBLAST, HYPOBLAST). Germ-layer of animal embryo, composed, like mesoderm, of cells which have moved from surface of embryo into its interior during gastrulation; developing into greater part of gut with its associated glands, etc. Term is usually applied to the germ-layer while it is still a demarcated region of the embryo, after gastrulation but before differentiation into derived tissues; and all tissues at later stages derived from the embryonic layer may be called endodermal. See *Germ-Layer, Ectoderm, Mesoderm.*

ENDODERMIS. Innermost layer of cortex, surrounding the stele. Characteristic of all roots, and of stems of pteridophytes and some dicotyledons. Consists of closely fitting cells with distinctive wall modifications. Most characteristic is band of wall material, *Casparian strip*, impervious to water, in radial and transverse walls. With age, especially in monocotyledons, cells (except *passage cells*) may become further modified by deposition of layer of suberin over entire inner surface of wall followed, particularly on inner tangential wall by layer of cellulose, sometimes lignified. Important physiologically in control of transfer of water and solutes between cortex and vascular cylinder. See *Casparian strip, Passage cell.*

ENDOMETRIUM. Glandular mucous membrane (q.v.) lining uterus of mammals. Undergoes cyclical growth and regression or destruction during sexual maturity. See *Oestrous Cycle, Placenta.*

ENDOMITOSIS. Doubling of chromosomes without division of nucleus, producing polyploidy (q.v.). Doubling may be repeated many times in a single nucleus. Occurs particularly in insects, where different adult tissues have characteristic degrees of polyploidy, and in some vertebrate and plant tissues.

ENDOPARASITE. Organism living parasitically within another. E.g. tapeworm, many Fungi. Cf. *Ectoparasite.*

ENDOPLASM. (Bot.). (*Endoplast*) Cytoplasm internal to plasma membrane. (Zool.). Internal cytoplasm of many cells, e.g. ova, Protozoa, which differs from ectoplasm (q.v.), usually in greater fluidity (in which case it may be called *plasmasol*) and in presence of many granules.

ENDOPLASMIC RETICULUM. System of membranes, too thin to be made visible except by electron microscopy, forming sheets and vesicles in the cytoplasm of many cells. Especially extensive in cells making large amounts of protein, where the membranes are covered with small granules (ribosomes) known to be concerned in protein synthesis.

ENDOPROCTA. Small group of sedentary aquatic animals sometimes regarded as a class of Polyzoa (q.v.), sometimes as a separate phylum.

ENDOPTERYGOTA (HOLOMETABOLA). Sub-class of insects in which larva differs strongly from adult and gives rise to it by a process of drastic change (metamorphosis), e.g. butterflies; as distinct from the Exopterygota in which the adult form develops more gradually.

END-ORGAN. Small organ composed of one or several cells, connected to the C.N.S. by a fibre of the peripheral nervous system. May be a *receptor* (q.v.) or may transform nerve-impulses into a stimulus to an effector, e.g. *motor end-plates* in striped muscle of vertebrates.

ENDOSKELETON. Skeleton lying inside the body, e.g. the bony skeleton of vertebrates. Cf. *Exoskeleton.* Arthropods, e.g. insects, although characterized by an exoskeleton may have an endoskeleton too; this is formed by folds of epidermis extending into the body and there secreting hardened cuticle.

ENDOSPERM. Nutritive issue surrounding and nourishing the embryo in seed plants, formed in embryo sac (q.v.) by division of endosperm nucleus after fertilization. In some seeds (*non-endospermic* or *exalbuminous*), endosperm is entirely absorbed by embryo by the time the seed is fully developed, e.g. pea and bean seeds; in other seeds (*endospermic* or *albuminous*), part of the endosperm remains and is not absorbed until the seed germinates, e.g. wheat, castor oil. Cf. *Perisperm.*

ENDOSTYLE. Structure concerned in ciliary feeding, present in Urochordata, Acrania, and ammocoete larva of lamprey. Consists of a ciliated and glandular groove or pocket in ventral wall of pharynx. Produces threads of mucus to which food particles adhere and which are passed round pharynx and backwards into gullet by ciliary action. Thyroid of adult lamprey develops from duct of endostyle of ammocoete larva. Thyroid of vertebrates is therefore considered to be homologous with endostyle.

ENDOTHELIUM. Single layer of smooth flattened cells lining heart, blood vessels, and lymph vessels, in vertebrates. See *Epithelium.*

ENDOTROPHIC. (Of mycorrhizas), with the mycelium of the fungus within cells of root cortex, e.g. in orchids. See *Mycorrhiza.* Cf. *Ectotrophic.*

ENTERIC CANAL. Alimentary canal (q.v.).

ENTEROKINASE. Enzyme (peptidase) secreted by vertebrate small intestine which converts trypsinogen to trypsin (q.v.).

ENTEROPNEUSTA. Hemichordata (q.v.).

ENTOMOGENOUS. (Of fungi), parasitic on insects.

ENTOMOLOGY. Study of insects.

ENTOMOPHILY. Pollination by insects. Cf. *Anemophily.*

ENVIRONMENT. Collective term for the conditions in which an organism lives, e.g. temperature, light, water, other organisms. Cf. *Internal environment.*

ENZYME. A protein which is a catalyst (i.e. a substance which in minute amounts promotes chemical change without itself being used up in the reaction), by virtue of its power of increasing the reactivity of a specific substance or specific substances (called the *Substrate*). There are many different kinds, each kind directly promoting only a very limited range of chemical reactions.

Within all living things a large number of chemical reactions are always in progress. See *Metabolism*. Most of the reactions of metabolism would not occur perceptibly in the absence of enzymes, at the temperature and in the other conditions in which living things exist; and so metabolism is entirely dependent on enzymes. An enzyme produces its effect on the substrate by combining with it and activating it, so that the substrate undergoes further chemical change, at the same time losing its combination with the enzyme. As the result of such a process, therefore, the enzyme is not consumed, but at the end of the process is free to deal with more substrate. Since the activation of the substrate is rapid, a very small amount of enzyme can produce a very great effect. One molecule of catalase can decompose 40,000 molecules of hydrogen peroxide per second at freezing point. Most enzymes are in fact present in relatively minute amounts.

Most enzymes activate only one kind of substrate each. Other enzymes are less highly specific, but even so each reacts only with chemically related substrates. There is correspondingly a large number of different enzymes, for all the different reactions of metabolism.

Each enzyme requires certain definite conditions for optimum performance, particularly as regards pH, the presence of specific accessory substances (co-enzymes, activators) and the absence of specific inhibiting substances.

Enzymes are unstable substances, and are easily destroyed or inactivated, e.g. by high temperature, or by a great range of chemical substances. It is probable that, although they are not consumed by the reaction, enzymes slowly break down and have to by synthesized. The micro-nutrients required by organisms probably mostly go to regenerating enzymes and co-enzymes.

Enzymes are all produced within living cells and most of them do their work there, though they can often be extracted and can be studied outside the cell. Some enzymes (e.g. digestive enzymes of vertebrates) are normally secreted to the outside of cells. In metabolic processes of cells, and also in digestion, enzymes work very much as systems, the products of one reaction becoming the substrate of another.

Closely similar enzymes, and enzyme systems, have been found in a wide variety of organisms, including plants, animals, and bacteria, and they account for the fundamental similarities of many aspects of their metabolism.

Several enzymes have been obtained pure and crystalline. Some of these are entirely protein (e.g. proteolytic enzymes); some are protein plus a prosthetic group (q.v.) (e.g. many concerned with oxidation and reduction). Enzymes are usually named by attaching the suffix *-ase* to a root indicating the nature of the substrate or of the reaction. See *Amylase*, *Catalase*, *Dehydrogenase*, *Proteolytic Enzymes*, *Adaptive Enzyme*.

EOCENE. Geological period (q.v.), subdivision of Tertiary, lasted approximately from 70 till 40 million years ago.

EOSINOPHIL LEUCOCYTE Polymorphonuclear leucocyte of vertebrates, containing granules staining in acid dyes such as eosin. In human beings normally about 2-5 per cent of all leucocytes, but become much increased in certain parasitic infections, and in allergies.

EPHEMERAL. Plant with short life-cycle (seed germination to seed production), having several generations in one year, e.g. groundsel. Cf. *Annual*, *Biennial*, *Perennial*.

EPHEMEROPTERA. Mayflies. Order of exopterygote insects, with aquatic nymphs which live a long time and may moult as many as twenty-three times; but adults live only from a few minutes up to a day and cannot eat or drink.

EPICOTYL. That portion of seedling stem above cotyledons.

EPIDEMIC. Large-scale temporary increase in prevalence of a disease due to a parasite.

EPIDERMIS. Outermost layer of cells of a plant or animal. In plants one cell-layer thick, covered in aerial parts by a non-cellular protective cuticle (q.v.). In vertebrates there is no non-cellular cuticle, and epidermis is several cell-layers thick; outermost ones being dead and horny (keratinized) in land-living vertebrates. See *Stratum Corneum*. In invertebrates the epidermis is only one cell-layer thick, and often secretes a cuticle.

EPIDIDYMIS. Long convoluted tube attached to testis of amniote vertebrates. Receives into one end sperm from testis tubules, which it stores and ripens. Other end leads into vas deferens and so to exterior. Embryonically derived from mesonephros and its duct.

EPIGAMIC CHARACTER. Character of an animal which is concerned in sexual reproduction; but other than gonads, and the ducts with their associated glands which convey the gametes. E.g. bird song.

EPIGEAL. (1) Of seed germination, seed leaves (cotyledons) appearing above ground, e.g. lettuce, tomato. Cf. *Hypogeal*. (2) Of animals, inhabiting exposed surface of land, as distinct from underground.

EPIGENESIS. Origin of entirely new structure during embryonic development. Cf. *Preformation*.

EPIGLOTTIS. Flap of mucous membrane and cartilage in mammals at base of tongue, on ventral wall of pharynx, against which glottis is pushed and thus closed when swallowing.

EPIGYNOUS. See *Receptacle*.

EPINASTY. (Bot.). More rapid growth of upper side of an organ, e.g. in a leaf, resulting in downward curling of leaf-blade. Cf. *Hyponasty*.

EPINEPHRINE. Adrenalin (q.v.).

EPIPETALOUS. (Of stamens), borne on the petals, with stalks (filaments) more or less fused with petals and appearing to originate from them.

EPIPHYSIS. (1) Separately ossified end of growing bone, forming the joint; peculiar to mammalian limb-bone and vertebra. Epiphysis is separated from rest of the bone (diaphysis, q.v.) by a cartilage plate (epiphyseal cartilage); and growth in length of the whole bone occurs by encroachment into this plate of new bone from diaphysis side, and formation of new cartilage on epiphysis side. When growth is complete, epiphysis and diaphysis fuse. (2) A synonym for pineal body (q.v.).

EPIPHYTE. Plant attached to another plant, not growing parasitically upon it but merely using it for support, e.g. various lichens, mosses, and orchids, epiphytes on trees.

EPITHELIUM. (Zool.). Sheet or tube of firmly coherent cells, with minimal material between the cells. (The term is, however, often not applied to such sheets or tubes when they have been embryonically derived from mesoderm, e.g. lining of blood-vessels (endothelium) or of coelom (mesothelium.) Lines cavities and tubes and covers exposed surfaces of body; one surface of epithelium being therefore free, other resting usually on connective tissue. Its cells are frequently secretory, and secretory part of most glands is made of epithelium. Epithelia are classified according to height of cell (dimension at right angles to extension of sheet) relative to breadth, into *columnar*, *cubical*, and *squamous* (in order of diminishing relative height) and according to whether sheet is one cell thick (simple) or many (stratified). Non-epithelial tissues are connective tissue, muscle, nervous tissue. See Fig. 2, p. 61. (Bot.). Layer of cells lining schizogenously formed secretory canals and cavities, e.g. in resin canals of pine.

EPIZOITE. Non-parasitic sedentary animal living attached to another animal.

EQUATORIAL PLATE. Arrangement of chromosomes during metaphase of mitosis (q.v.) or meiosis (q.v.) when all lie approximately in one plane, at equator of spindle (q.v.).

EQUISETALES. Horsetails. Order of Pteridophyta with fossil record extending back to Palaeozoic. Attained their highest development in Carboniferous and included tree-like forms, e.g. *Calamites*. Now represented by single genus *Equisetum*. Perennial, herbaceous plants with rhizome sending up erect, grooved stems bearing at nodes whorls of rudimentary scale-leaves. Stem unbranched or bearing whorls of slender branches in axils of scale-leaves. Stem green, photosynthetic. Sporophylls compacted into distinct cones arising from cone axis at right angles and bearing several sporangia on under surface of terminal shield-shaped expansion. Homosporous, prothalli monoecious or dioecious depending on environmental conditions.

EREPSIN. Mixture of proteolytic enzymes produced by wall of small intestine of vertebrates; mainly peptidases (q.v.).

ERGOT. (1) Disease of inflorescence of cereals and wild grasses caused by the ascomycete fungus *Claviceps purpurea*. (2) Dark spur-shaped sclerotium developing in place of a healthy grain in a diseased inflorescence. Ergots contain substances poisonous to man and animals. Some (ergotamine) are used in medicine.

ERYTHROCYTE. Red blood corpuscle (q.v.).

ESCAPE. Cultivated plant found growing as though wild.

ETAERIO (Of fruits) an aggregation, e.g. of achenes, in buttercup; of drupes, in blackberry.

ETHOLOGY. Study of behaviour of an animal in its normal environment.

ETIOLATION. Phenomenon exhibited by green plants when grown in darkness. Such plants are pale yellow because of absence of chlorophyll, their stems are exceptionally long owing to abnormal lengthening of internodes, and their leaves reduced in size.

EUCHROMATIN. Chromosome material showing its maximum staining during metaphase (q.v.) and less dense staining in the resting nucleus. Contains the genes of major effect. Cf. *Heterochromatin.*

EUGENICS. Study of possibility of improving humanity by altering its genetic composition by encouraging breeding of those presumed to have desirable genes, and discouraging breeding of those presumed to have undesirable genes. It is rarely known who has which.

EUPHOTIC ZONE. Zone near surface of sea (roughly upper 100 metres) into which sufficient light penetrates for active photosynthesis.

EUPLOID. Having each of the different chromosomes of the set present in the same number; therefore with an exact multiple of the haploid chromosome number, e.g. diploid, polyploid, Cf. *Aneuploid.*

EURYHALINE. Able to tolerate wide variation of osmotic pressure of environment. Cf. *Stenohaline.*

EURYPTERIDA. An extinct (Palaeozoic) order of Arachnida, resembling scorpions in some features, but aquatic; some were very large (six feet long).

EURYTHERMOUS. Able to tolerate wide variations of temperature of environment. Cf. *Stenothermous.*

EUSTACHIAN TUBE. Tube connecting middle ear to pharynx in

tetrapod vertebrates. Allows air pressure on inner side of ear drum, i.e. in middle ear, to be equalized with that outside the animal, preventing distortion of ear-drum. See *Spiracle*.

EUTHERIA. Placentalia (q.v.).

EVERGREEN. Bearing leaves all the year round, e.g. laurel, pine. Cf. *Deciduous*.

EVOCATION. Induction in an embryonic tissue of a particular process of development, by a chemical substance (an *evocator*) diffusing from nearby tissue (or from an artificial implant). Best-known instance is evocation in vertebrates of neural development in ectoderm, as a result of diffusion of (an unknown) substance from underlying chorda-mesoderm in gastrula stage. See *Organizer*.

EVOLUTION. Cumulative change in the characteristics of populations of organisms, occurring in the course of successive generations related by descent. The theory that evolution accounts for origin of all kinds of organisms now existing is opposed to the theory of special creation, i.e. that each kind of organism was created as such and is therefore not related by descent to any other. See *Natural Selection*.

EXCRETION. Getting rid of products of metabolism (either by storing them in insoluble form, or by removing them from body). In animals particularly applies to products of protein metabolism, organs mainly concerned being kidneys of vertebrates, Malpighian tubes of insects, and probably nephridia in many other invertebrates.

EXODERMIS. Layer of closely fitting cortical cells with suberized walls, replacing the withered piliferous layer in older part of roots.

EXOPTERYGOTA (HETEROMETABOLA, HEMIMETABOLA). Subclass of insects in which adult form develops gradually by successive moults (cockroaches, mayflies, etc.) as distinct from the Endopterygota which have rapid metamorphosis.

EXOSKELETON. Skeleton covering the outside of the body, or situated in the skin. In Arthropoda the exoskeleton, e.g. cuticle (q.v.) of insects, or lime-impregnated cuticle of crabs, is superficial to the epidermis. Shell of many molluscs is a similar exoskeleton of lime-impregnated cuticle. In many vertebrates, e.g. scaly fish, tortoises, armadillo, the exoskeleton consists chiefly of bony plates beneath epidermis; in tortoise and armadillo these plates have a horny covering of keratin (q.v.) derived from epidermis.

EXPLANTATION. Tissue culture (q.v.).

EXPRESSIVITY. Intensity with which effect of a gene is realized in the phenotype. Influenced by genotype and environment to varying degrees. See *Penetrance*.

EXSICCATA (often abbreviated to EXSIC.). Dried, preserved plant material deposited in a herbarium.

EXTEROCEPTOR. Receptor (q.v.) which detects stimuli emanating from outside the animal, e.g. light, heat, sound. Cf. *Interoceptor*.

EXTRACELLULAR. (Zool.). Within an organism, but not within its constituent cells. Cf. *Intracellular.*

EXTRAEMBRYONIC. (Zool.). Derived from the zygote, but lying outside epidermis of embryo proper. *E. membranes* are those extraembryonic structures of vertebrates concerned in nutrition, respiration, and protection of embryo: yolk-sac, amnion, chorion, allantois.

EXTRORSE. (Of the dehiscence of an anther), away from centre of flower. Cf. *Introrse.*

EYE. Receptor organ for light. Some form of eye occurs very widely among animals, varying from the simple ocellus to the very com-

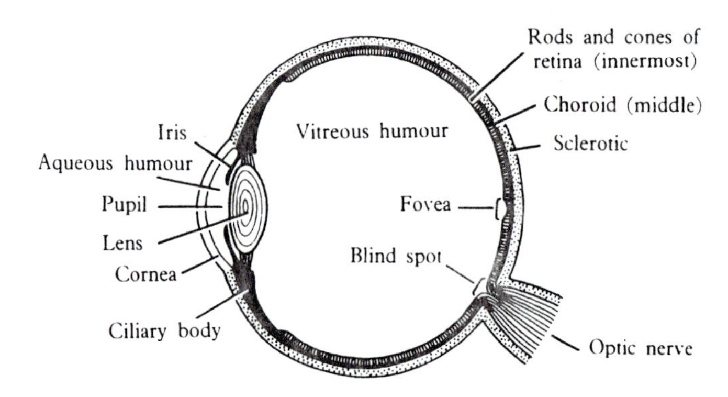

Fig. 3. Section of mammalian eye through lens and optic nerve.

plex organs of insects (see *Eye, Compound*), vertebrates (see Fig. 3) and cephalopod molluscs. Eyes of two latter groups are extraordinarily similar, offering an interesting example of analogous (q.v.) organs. See parts labelled in Fig. 3, *Accommodation, Binocular Vision.*

EYE, COMPOUND. Eye of insects and Crustacea. Consists of a number of separate elements (ommatidia) each with light-sensitive cells and refractive system which can form an image; separated from each other in varying degrees by pigment. According to the structure of the eye, and to the distribution of pigment between the ommatidia, the eye can form either *apposition* images, in which case each ommatidium will focus only rays almost parallel to its long axis, so that each forms an image of only a very small part of the visual field, an image of the whole resulting from combination of these part images; or *superposition* images in which the sense cells of an ommatidium may receive light from a large part of the visual field so that the image received may overlap those received by as many as thirty neighbouring ommatidia. The superposition image thus gains in luminosity but loses in sharpness compared with the apposition image. Diurnal insects have apposition eyes, nocturnal ones superposition, but there are many intermediate grades; and in some cases one may change temporarily into the other by movement of pigment between the ommatidia (dark adaptation).

EYE-MUSCLES. (a) *Extrinsic Eye-muscles.* Those outside eyeball. The set of six muscles which move each eyeball, characteristic of all vertebrates with functional eyes. Consists of anterior (i.e. near to nose) pair of oblique muscles and four, more posterior (i.e. nearer to ears), rectus muscles. Supplied by third, fourth, and sixth cranial nerves. (See *Oculomotor, Trochlear, Abducens, Nerves.*) (b) Those inside eyeball. *Intrinsic Eye-Muscles.* See *Iris, Ciliary Body.*

F

F_1 (FIRST FILIAL GENERATION). The offspring resulting from crossing the plants or animals of parental generation (P_1) from which an experiment starts.

F_2 (SECOND FILIAL GENERATION). The offspring resulting from crossing the members of the F_1 (q.v.) among themselves.

FACIAL NERVE. Seventh cranial nerve of vertebrates. See *Hyoid Arch.* In mammals mainly motor, going to superficial muscles of face, salivary glands; also to taste-buds of front of tongue. A dorsal root (q.v.).

FACILITATION. Increase in the responsiveness of a nerve-cell or of an effector cell produced by summation (q.v.) of impinging impulses.

FACTOR (HEREDITARY). Gene (q.v.).

FAECES. Indigestible residue of food, together with residue of secretions, bacteria, etc., expelled from alimentary canal through anus.

FALLOPIAN TUBE (UTERINE TUBE). In female mammals, tube, with funnel-shaped opening just beside ovary, leading from peritoneal cavity to uterus. One on each side. By muscular and ciliary action it conducts eggs from ovary to uterus, and sperms from uterus to that place in upper part of tube where they fertilize descending eggs. Represents part of Müllerian duct of other vertebrates.

FAMILY. One of the kinds of group used in classifying organisms. Consists of a number of similar genera (sometimes of only one genus). Similar families are grouped in an order. In zoology, name of a family ends in – *idae*; in botany usually in – *ceae.* Families of flowering plants used to be called *natural orders.* See *Classification.*

FASCIA. Connective tissue; applied particularly to sheets thereof. E.g. *superficial fascia* of mammals, the loose connective tissue beneath dermis, containing fat in man and whale (blubber); *deep fascia*, tough sheets enclosing muscles and groups of muscles.

FASCIATION. Coalescing of stems, branches, etc., to form abnormally thick growths.

FASCICULAR (INTRAFASCICULAR) CAMBIUM. Cambium (q.v.) arising within a vascular bundle. Cf. *Interfascicular cambium*.

FAT. (1) Substance extractable from tissue by 'fat-solvent' (such as ether or hot alcohol) and containing a fatty acid. Lipide, Lipid, Lipin and Lipoid have all been used as synonyms for fat in this sense; but they also have other (often ambiguous) more restricted meanings. (2) In biochemical analysis 'fat' may refer simply to any substance extractable in fat solvent; and besides those containing fatty acids, would therefore comprise sterols (which however often occur chemically combined with fatty acids), steroids, carotene and, in plants, terpenes. (3) *True* (or *neutral*) *fat* is compound of glycerol and fatty acids.

FAT-BODY. (1) In many Amphibia and lizards, organ in abdomen containing fat as adipose tissue used up during hibernation (2) In insects, diffuse tissue beneath skin, around gut, etc., storing fat, proteins, and sometimes glycogen and uric acid.

FATTY ACID. Organic aliphatic acid. Biological ones have usually straight chains and an even number of carbon atoms.

FAUNA. The animal population present in a certain place or at a certain epoch.

FEMUR. Of tetrapod vertebrates, the thigh-bone. See Fig. 7, p. 175. Of insects, one of the segments (third from base) of the leg.

FERMENTATION. Decomposition of organic substances by organisms, especially bacteria and yeasts. E.g. decomposition of sugar forming ethyl alcohol and carbon dioxide by yeast in making of wine; or of ethyl alcohol forming acetic acid in making of vinegar. Not usually applied to decomposition of proteins, which is called putrefaction. Sometimes means anaerobic respiration.

FERN. See *Filicales*.

FERTILIZATION. The union of two special cells, the gametes (q.v.), the essential process of sexual reproduction. The resulting single cell is a zygote. Fertilization has two aspects: (*a*) fusion of two haploid nuclei which brings together in the zygote a selection of genes (q.v.) from two distinct individuals, i.e. the two parents; (*b*) initiation of development of a new individual. In parthenogenesis (q.v.) and pseudogamy (q.v.) aspect (*b*) occurs without aspect (*a*). *External fertilization*, union of gametes outside body of parents, as in many aquatic animals, e.g. Echinodermata. *Internal fertilization*, union of the gametes inside the female, as in most terrestrial animals (insects, tetrapods) and many aquatic animals of many different phyla. *Fertilization membrane*. Tough membrane which appears at surface, or separates from surface, of many eggs when fertilized; in most cases it is the same as the vitelline membrane (q.v.).

FEULGEN METHOD. Method applied to histological sections, giving purple colour where there is desoxyribosenucleic acid (e.g. in chromosomes).

FIBRE. (Bot.). Element of sclerenchyma (q.v.); (Zool.). See *Collagen, Reticulin, Elastin.*

FIBRIN, FIBRINOGEN. See *Blood clotting.*

FIBROBLAST (FIBROCYTE). Kind of cell of irregular, branching shape, found distributed throughout vertebrate connective tissue. Function is apparently to form and maintain collagen. Similar shaped cell occurs in many invertebrates, possibly of same function.

FIBROUS ROOT. A fibrous root system consists of a tuft of adventitious roots of more or less equal diameter, arising from stem base or hypocotyl, and bearing smaller lateral roots, e.g. wheat, strawberry. Cf. *Tap root.*

FIBULA. The posterior of the two bones (other is tibia) in shank (below the knee) of hind-limb of tetrapod vertebrate. Lateral bone in lower leg of man. See Fig. 7, p. 175.

FILICALES. Ferns. Order of Pteridophyta including great majority of existing pteridophytes and a few extinct forms. Perennial plants with creeping or erect rhizome, or with erect aerial stem several metres in height, e.g. tropical tree ferns. Leaves characteristically large and conspicuous. Sporophylls like ordinary vegetative leaves, bearing sporangia on under surface, often in groups (sori), e.g. shield fern, or much modified, superficially unlike leaves, e.g. royal fern. Homosporous, prothalli bearing both antheridia and archegonia. Includes also a small group of aquatic, heterosporous ferns.

FIN. Of aquatic vertebrates: *Median f.* in mid-line of body, unpaired, either continuous, or as separate *dorsal*, *caudal* (which functions in propulsion of fish and Cetacea) or *ventral* fins. *Paired* fins are lateral: *pectoral* attached to shoulder-girdle, *pelvic* attached to pelvic girdle; these two pairs, homologues of tetrapod limbs though they do not contain comparable bones, occur in most fish (see however *Acanthodians*) but not in Agnatha; their usual function is control of direction of movement.

FIN-RAYS. Supports for fins; in most fish radiating skeletal rods, either cartilage, bone or collagen-like (elastoidin).

FISH. Pisces (q.v.).

FISSION. Reproduction by splitting into two equal parts (*binary fission*) or more than two equal parts (*multiple fission*). Binary fission occurs in many unicellular organisms, and some multicellular ones (e.g. corals); multiple fission occurs in the course of spore formation by Protozoa. Cf. *Budding.*

FIXATION. The first step in making permanent preparations of organisms or tissues for microscopic study: aims at killing cells, i.e. preventing subsequent change through metabolism or decay, with the least distortion of structure and in such a way that subsequent study is as easy as possible. Various chemicals are used as fixatives (e.g. formaldehyde, osmium tetroxide), mainly as mixtures. Almost all produce some artefacts, particularly a coarse precipitation within nucleus and cytoplasm.

FLAGELLATA (MASTIGOPHORA). Class of Protozoa characterized by possession of one or more flagella. Includes both plant- and animal-like forms and some with mixed characteristics. Opinions differ as to the position of some flagellated unicellular plants with chlorophyll and cellulose wall; some biologists include them in the Flagellata and some place them in the Algae. Different groups of Flagellata may be related to the various ancestors of the Metazoa, Parazoa, and Metaphyta, and it is possible that the other Protozoan groups were derived from flagellated forms.

FLAGELLUM. Fine long thread having lashing or undulating movement, projecting from a cell. Has an internal structure, shown by electron microscope, consisting usually of nine very fine longitudinal fibrils arranged to form a cylinder, with another pair of fibrils lying within this cylinder in the central axis. Very widely distributed; characteristic of one group of unicellular organisms (Flagellata), present in most motile gametes, in zoospores, and in some metazoan cells, e.g. endodermal cells of *Hydra*. Usually only one or a few per cell, though occasionally, in some Protozoa, many. Motion of most flagella probably consists of waves which travel along them. Flagella are responsible for movement of those unicellular organisms and reproductive cells which bear them; and they move water through body of sponges. Bacterial flagella are rather different, much smaller in diameter and without the internal structure of other flagella. See *Choanocyte*. Cf. *Cilium*.

FLAME CELL (SOLENOCYTE). Hollow cup-shaped cell with a bunch of cilia which work in the lumen. Flame cells are usually connected together by canals which ultimately open to the exterior. See *Nephridium*. Occur in several groups: Platyhelminthes, Nemertea, Rotifera, Annelida, larvae of Molluscs, amphioxus.

FLATWORMS. Playhelminthes (q.v.).

FLORA. (1) Plant population of a particular area. (2) List of plant species (with descriptions) of a particular area arranged in families and genera, together with a key to aid identification.

FLORAL DIAGRAM. Diagram illustrating relative position and number of parts in each of the sets of organs comprising a flower. See Fig. 4, p. 91, and *Floral Formula*.

FLORAL FORMULA. An expression summarizing the information given in a floral diagram (q.v.). The floral formula of buttercup, $K_5C_5A_\infty G_{\underline{\infty}}$, indicates a flower with a calyx (K) of five free sepals, corolla (C) of five free petals, androecium (A) of an indefinite number of stamens and a gynoecium (G) of an indefinite number of free carpels. The line below the number of carpels indicates that the gynoecium is superior. The floral formula of the campanula,

$K_5C_{(5)}A_5G(\bar{5})$, shows that the flower has five free sepals, five petals united ( ) to form a gamopetalous corolla, five stamens, and five carpels united ( ) to form a syncarpous gynoecium. The line above

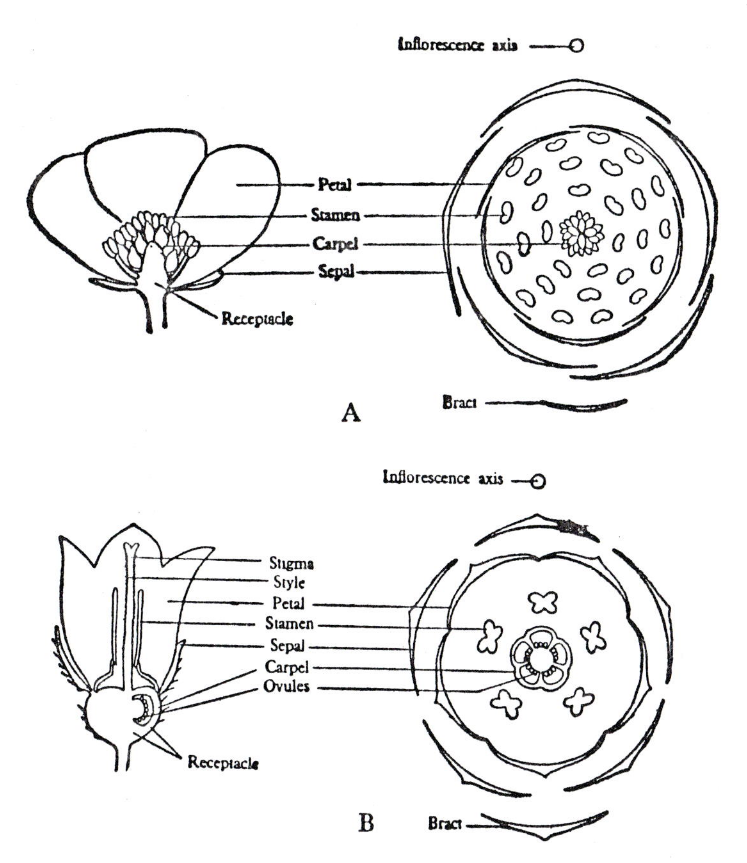

Fig. 4. Diagrams illustrating flower structure; median vertical section of flower (left) and floral diagram (right). A. Buttercup; B. Campanula.

the carpel number indicates that the gynoecium is inferior. See Fig. 4, above.

FLOWER. Specialized reproductive shoot, consisting of an axis (*receptacle*), on which are inserted four different sets of organs. Outermost *sepals*, usually green, leaf-like, in the bud stage enclosing and protecting the other flower parts, and collectively known as the *calyx*. Within the sepals are the *petals*, usually conspicuous, brightly coloured parts, collectively known as the *corolla*. Calyx and corolla together constitute the *perianth*; they are not directly concerned in reproduction and are often referred to as *accessory* flower parts. Within the petals are the *stamens* (*microsporophylls*), each consisting of a stalk bearing an *anther*, in which *pollen grains* (*microspores*) are produced. In the centre of the flower is the *gynoecium*, comprising one or more *carpels* (*megasporophylls*), each consisting of an *ovary*, a terminal prolongation the *style*, bearing the *stigma*, receptive surface for pollen grains. The ovary contains a varying number of *ovules* which after fertilization develop into seeds. Stamens and carpels are together known as *essential* flower parts since they alone are concerned in reproduction. See Fig. 4, p. 91.

FLUKE. Parasitic flatworm (class Trematoda). Adults live in liver, gut, lung, or blood-vessels of a vertebrate, where they may cause serious disease (e.g. liver fluke; *Schistosoma*). Larvae (*miracidia*) which emerge from eggs passed out with excreta of host, parasitize snails, in which they reproduce asexually, giving rise to *rediae*, which ultimately produce *cercaria* (q.v.) larvae. These are immature flukes and are infective to another host, which may be the vertebrate in which the fluke becomes sexually mature, or in some cases another intermediate host.

FOETAL MEMBRANES. Extraembryonic membranes (q.v.) of mammal.

FOETUS. Mammalian embryo after recognizable appearance of main features of fully developed animal; in man after about two months of gestation.

FOLLICLE. (Bot.). Dry fruit formed from a single carpel which splits open along a single line of dehiscence to liberate its seeds, e.g. larkspur, columbine. (Zool.). See *Ovarian Follicle*, *Graafian Follicle*, *Hair-Follicle*.

FOLLICLE-STIMULATING HORMONE (F.S.H.). Gonadotrophic hormone (q.v.) secreted by anterior lobe of pituitary in mammals, which stimulates growth of Graafian follicles and oestrogen production of ovary, and development of spermatozoa in testis.

FOLLICULAR PHASE. See *Oestrous Cycle*.

FONTANELLE. A gap in the skeletal covering of the brain, either in the chondrocranium, or between the dermal bones. The gap is covered only by skin and fascia. In human baby there is conspicuous parietal fontanelle on top of head between frontal and parietal bones.

FOOD-CHAIN. Metaphorical chain of organisms existing in any natural community such that each link in the chain feeds on the one below and is eaten by the one above. The number of links seldom exceeds six. At the bottom of the chain are plants, bacteria, or scavenging forms; at the top, the largest carnivores. Proceeding up the chain, animals get larger but fewer. All the food-chains in a community make up the *food-cycle*.

FORAMEN. Opening. *F. magnum*, opening in back of vertebrate skull at articulation with vertebral column, through which spinal cord passes. *F. ovale*, of heart; opening between right and left auricles of embryonic mammal, allowing a large part of the oxygenated blood, which returns from the placenta through the vena cava inferior to the right auricle, to go straight to the left auricle, then to left ventricle and so through the aorta to the arteries of the body, instead of to the right ventricle and thence to pulmonary artery. Normally closes at birth.

FORAMINIFERA. Group of rhizopod Protozoa, mostly marine, which form shells usually of lime. Some shells are microscopic, but others, colonial forms, may be as big as a halfpenny. Foraminiferan shells form an important part of chalk, and of many deep-sea oozes (see *Globigerina Ooze*).

FORE-BRAIN (PROSENCEPHALON). Anterior of the three divisions marked out in the embryonic brain of vertebrates by constrictions. Gives rise during development to cerebral hemispheres, eyes, hypothalamus, etc.

FORM. (Bot.). Smallest of the groups used in classifying plants. Category within species generally applied to members showing trivial variations from type, e.g. in colour of corolla. (Zool.). One of the kinds of a polymorphic species (see *Polymorphism*); a seasonal variant; or used as a neutral term in taxonomy when it is unclear as yet whether a species or sub-species or some minor grouping is the appropriate classification.

FOSSIL. Remains of organism, or direct evidence of its presence, preserved in rocks. Generally hard parts only are preserved, usually partly or wholly replaced by minerals deposited from circulating water, as impressions or casts.

FOVEA. Shallow pit in retina of some vertebrates, which is place of greatest acuity of vision (Fig. 3, p. 86). Contains no rods, but very numerous cones; and there are no blood-vessels, and no thick layer of nerve fibres, interposed between cones and incoming light, as in rest of retina. Occurs in diurnal birds and lizards, and in primates including man. An area of relatively acute vision, or *macula*, without a fovea, is found in many vertebrates (area containing yellow pigment surrounding fovea of man and some primates is *macula lutea*). In binocular vision the two eye-balls are orientated so that image of one object falls on both their foveae (or maculae); this object is 'looked at', i.e. it has predominant part in visual stimulation of the animal. Many vertebrates without binocular vision also similarly 'look at' objects with fovea or macula of one eye.

FRATERNAL TWINS. Dizygotic twins (q.v.).

FREE-MARTIN. Female member of unlike-sexed twins in cattle and occasionally in other ungulates. Sterile, and partially converted towards hermaphrodite condition by some influence (probably hormonal) of its twin brother, which reaches the free-martin because placental circulations fuse.

FRONTAL BONE. Large dermal bone covering front part of brain in vertebrates (region of forehead in man). Air-spaces extend from nasal cavity into frontal bones of mammals (*frontal sinuses*).

FRUCTOSE (LAEVULOSE). A 6-carbon-atom sugar (hexose). Combined with glucose in sucrose. Widely distributed in plants.

FRUIT. Ripened ovary of the flower, enclosing seeds.

FRUSTULE. Diatom cell; wall consists of two halves or valves, the older (*epitheca*) fitting closely over the younger (*hypotheca*).

FUCOXANTHIN. Pigment present with chlorophyll, in brown Algae.

FUNCTION. The function of part of an organism is the way in which that part helps maintain the organism to which it belongs alive and able to reproduce; or sometimes it means simply the way it works, the processes going on in it. See *Adaptation* (Evolutionary).

FUNGI. Mushrooms, moulds, rusts, yeasts, etc. Sub-division of Thallophyta. Fungi are simply-organized plants, unicellular or made of cellular filaments, *hyphae*, lacking green colouring matter (chlorophyll); reproducing asexually and sexually with the formation of spores, often produced in enormous numbers. Many fungi are microscopic, some, especially their fruit-bodies, e.g. mushroom, puff-ball, reach a fair size. Lacking chlorophyll fungi live either as saprophytes or as parasites of other plants and of animals. They are extremely important as agents of plant disease (they also cause a few animal diseases), and in the decay of food, fabrics, and timber. In soil they take part with other organisms in decomposition of plant and animal residues. Various fungi are used in industrial processes, e.g. brewing and baking, others provide valuable sources of certain food proteins and vitamins. The development of antibiotics from fungi, particularly of penicillin from *Penicillium notatum*, has been responsible for important advances in medical practice.

Fungi are divided, primarily on the type of sexual spore produced, into four classes; (1) *Phycomycetes*, possessing hyphae without cross walls, reproducing asexually by zoospores, aplanospores or conidia, and sexually with formation of thick-walled resting spores; (2) *Ascomycetes*, with hyphae divided into cells by cross septa (septate), with asexual reproduction by conidia and sexual reproduction with formation of *ascospores* in an almost spherical or cylindrical cell, the *ascus*. In most Ascomycetes the asci are grouped within visible fruit-bodies (cleistocarp, apothecium, perithecium); (3) *Basidiomycetes*, with septate hyphae and sexually produced spores, *basidiospores*, borne externally on a club-shaped or cylindrical cell, the *basidium*. In certain forms the basidia are grouped together in highly organized fruit-bodies, e.g. mushroom, puff-ball, stinkhorn, bracket-fungus; (4) *Fungi Imperfecti*, a group of Fungi that lack a sexually reproducing stage, thought to be mostly asexual forms of Ascomycetes in which sexual stage has been lost during evolution or has not yet been identified.

FUNGICIDE. Substance destructive to fungi.

FUNICLE. Stalk of ovule, attaching it to placenta (q.v.).

G

GALACTOSE. A hexose sugar; constituent of lactose, and commonly of plant polysaccharides (many gums, mucilages, and pectins).

GALL-BLADDER. Small bladder in many vertebrates arising from bile duct (q.v.) near or in liver, storing bile between meals. Contractile walls empty bladder when there is food, especially fat, in intestine, probably activated to do so by hormones secreted by intestine wall.

GAMETANGIUM. (Bot.). Organ in which gametes are produced.

GAMETE (GERM-CELL). Reproductive cell whose nucleus and often cytoplasm fuses with that of another gamete (constituting fertilization), the resulting cell (zygote) developing into a new individual. Gametes are haploid. They are usually differentiated into female gamete (immobile, with massive cytoplasm, able to develop when stimulated, called the ovum) and male gamete (moves or is moved to female gamete, with cytoplasm reduced, e.g. sperm of animals), one of each fusing to form the zygote. See *Heterogamy*, *Isogamy*, *Sperm*, *Ovum*, *Fertilization*. Some animal and plant species produce cells, which are also called gametes, of substantially the same kind as a female gamete of the same or related organism, but which develop into a new individual without fertilization (i.e. by parthenogenesis, q.v.); such gametes are usually diploid.

GAMETOCYTE. Cell which undergoes meiosis, forming gametes. See *Oocyte; Spermatocyte.*

GAMETOPHYTE. Phase of life cycle of plants which has haploid (q.v.) nuclei; during it the sex-cells are produced. It arises from a (haploid) spore produced by meiosis from a (diploid) sporophyte (q.v.). See *Alternation of Generations.*

GAMOPETALOUS (SYMPETALOUS). (Of a flower), with united petals, e.g. primrose. Cf. *Polypetalous.*

GAMOSEPALOUS. (Of a flower), with united sepals, e.g. primrose. Cf. *Polysepalous.*

GANGLION. Small solid mass of nervous tissue containing numerous cell-bodies (q.v.). The central nervous system of many invertebrates consists largely of such ganglia, connected by nerve cords, very variously arranged but usually well-developed in the head (*cerebral ganglia*). In vertebrates the central nervous system has a different structure, but ganglia occur in the peripheral nervous system (see *Nerve-Root*) and autonomic nervous system. Some of the 'nuclei' of the brain are called ganglia.

GANOID. A primitive actinopterygian fish, e.g. *Acipenser*, *Polypterus.*

GANOID SCALE. Scale typical of primitive Actinopterygii. Same components as cosmoid scale (q.v.), but much thicker layer of superficial enamel-like substance (ganoine). Unlike cosmoid scale, grows in thickness by addition of layers of material all round, ganoine above as well as laminated bone below.

GASTRIC. (Adj.). Of the stomach.

GASTROPODA. Class of Mollusca, including snails, slugs, sea-hares. Marine, freshwater, and terrestrial. Head distinct, with eyes and tentacles; often a single (univalve) shell.

GASTRULA. Stage of embryonic development of an animal, succeeding blastula, when the gastrulation movements occur.

GASTRULATION. Embryological term for the complex of cell movements which occurs in almost all animals at the end of the cleavage period. See *Blastula*. The movements carry those cells whose descendants will form the future internal organs from their largely superficial position in the blastula to approximately their definitive positions inside the embryo. In many embryos, these presumptive internal organs undergo gastrulation in two main groups of cells, endoderm (future gut, etc.), and mesoderm (future muscle, blood, etc.). The ectoderm (future external layer, and also the future nervous system, which later becomes internal) is also rearranged during the gastrulation movements to approximately its definitive position. See *Blastopore*.

GEMMA. Organ of vegetative reproduction in mosses and liverworts; consisting of a small group of cells of varying shape and size that becomes detached from the parent and develops into a new plant; often formed in groups, in receptacles known as gemma-cups.

GEMMATION. Kind of asexual reproduction by formation of a group of cells (a gemma in plants) which develops into a new individual, or a new member of a colony of connected individuals. It may develop before or after it separates (completely or partially) from its parent. Occurs in many liverworts, mosses, coelenterates, ascidians. Commonly called *budding* in animals.

GEMMULE. Of sponges, a bud formed internally as a group of cells, which may become free by decay of parent and subsequently form a new individual. Freshwater sponges over-winter in this way.

GENE (HEREDITARY FACTOR). Unit of the material of inheritance. The properties of this material will be first discussed before considering in what sense a gene is a unit of it.

The material of inheritance. Much of the similarity between related organisms, especially the similarity on which taxonomy is based, is due to their possessing similar inherited material. Each individual receives a set of this material from its parent or parents; the material is duplicated with great exactness (the process of *replication*) during subsequent growth or preparation for reproduction; and, largely unchanged in structure, though with certain changes mentioned below (mutation, segregation), it is handed on to the individual's progeny. The material influences the characteristics developed by the individual so that the similarity between related organisms is produced. It is highly probable that in bacteria and many viruses the material of inheritance consists of deoxyribosenucleic acid (DNA), in other viruses ribosenucleic acid (RNA).

In plants and animals, the material is situated in the chromosomes of the nucleus (apart from a few instances of inheritance via *plasmagenes* in the cytoplasm). The chromosomes contain much protein as well as DNA, but the latter is now generally thought to be the hereditary material in these organisms.

The gene as a unit. The inherited material occasionally undergoes a change, known as *mutation*, the changed form (*mutant*) being duplicated and inherited instead of the original form. Such a mutant produces a change in certain characteristics of those organisms that inherit it. Many such mutants can be found in any large population of individuals of a species. Some of these mutants differ from each other in that each affects a different set of characteristics of the organism; and when, as is possible by various methods, the approximate position in the chromosome complex of these different mutations has been discovered, each is found to have its own distinct place or *locus*. Other mutants will also be found, in different members of the population, which (tested in constant conditions) affect the *same* set of characteristics in each organism, though in different ways; and these mutations turn out to have occurred in nearly or quite the same relative position in the chromosome complex of each of the organisms bearing them.

A *gene* is one particular piece of a chromosome (or of its contained hereditary material) influencing a particular set of characteristics which mutates as a whole. (A gene thus defined as part of a chromosome with unit function is coming to be called a *cistron*.)

Gene function. The question arises as to how there can be such units of function. This is at present answered by the 'one gene – one protein' theory, by which it is supposed that each gene is responsible for the production of one particular kind of protein, or more precisely for the exact sequence of amino-acids making up one specific kind of polypeptide chain (a protein molecule contains one or more polypeptide chains). All the other effects of genes can be derived from this, because enzymes are proteins, and enzymes influence substantially all the process that go on in organisms.

The hereditary material is then made up of genes. Each gene during growth can be repeatedly exactly duplicated, handing on its own structure to the copies. When a gene mutates, it similarly hands on its changed structure to its copies, and, according to the one gene – one protein theory, also makes a protein with a slightly different amino acid sequence. The mutant gene and the original gene are said to be *alleles* of (or *allelomorphic* to) each other. The detectable differences between organisms differing by having one allele substituted for another (whether the one gene – one protein theory is true or not) are usually distant repercussions of the primary difference. Consequently, in complex organisms, differences in numerous organs are produced by substitution of one allele for another, though some organs are more strikingly affected than others; and every organ can be affected in various ways by substitution at many different loci, i.e. is simultaneously influenced by many different genes. Environmental variation is likely to be simultaneously influencing the characteristics that the genes are influencing. Hence the effect of a given gene has to be assessed in standard conditions.

Genes and chromosomes. Every cell of each individual of each plant and animal species has commonly an identical set of chromosomes (of a number usually in double figures). Different chromosomes have arrays of different genes. Every cell of each individual has commonly likewise an identical set of genes (perhaps 5–10,000 in *Drosophila*). Individuals of the same species differ in having different members of some of the sets of alleles.

Each chromosome contains numerous genes arranged in single file ('linear order') along its length. The 'chromosomes' of bacteria and viruses also have this linear order. These genes are all different and none is an allele of any other. Every specimen of a particular chromosome has its genes arranged in the same characteristic order along its length. The fixed position in the chromosome (relative to other genes) always occupied by a given kind of gene is called a *locus*. Alleles have the same locus, though it must be emphasized that only one member of a set of genes that are alleles to each other (an allelomorphic series) can be present in the locus, and therefore in a given chromosome, at the same time.

Before nuclear division by mitosis (q.v.) the genes in each chromosome are duplicated, and the whole chromosome becomes double. Each of the two cells resulting from mitosis receives one of the duplicates and has therefore exactly the same complement of genes and chromosomes.

In the body-cells of animals and higher plants chromosomes occur in pairs, the members of which have identical appearance (except sex-chromosomes, q.v.). The members of such a pair have the same arrangement of loci; and the genes in the two corresponding loci of a pair of chromosomes may either be identical, or two different members of an allelomorphic series. If identical, the individual is said to be *homozygous* for that locus; if alleles, *heterozygous*.

Genes and the variety of individuals. While every body-cell of a multicellular organism, with rare exceptions, has an identical set of genes, this is not true of the gametes produced by an individual, which are of enormous variety. A gamete, as the result of meiosis (q.v.), has only one member of each pair of chromosomes. But the two members of any pair of chromosomes usually differ from each other, because some of the loci are heterozygous, so that gametes receiving different members of a pair of chromosomes will be different. Since it is a matter of chance which member of each pair of chromosomes a gamete receives, a gamete may with equal probability have any of the possible combinations of one member of each pair of chromosomes, i.e. any of 2^x combinations where x is the number of chromosome pairs. This chance distribution is known as *independent assortment.* The number of different gametes produced by an individual is actually much further increased, by the phenomenon of *crossing-over* (q.v.), the resulting chromosomes being different, in the particular set of alleles they contain, from any possessed by the parent organism. The combined effect of independent assortment and crossing over is called *segregation.* A further, but much rarer, source of diversity in gametes is through mutation during their formation.

Knowing the laws governing the distribution of the chromosomes to the gametes, and assuming that it is a matter of chance which male gamete fertilizes which female gamete, we can calculate the probable distribution of one or more alleles amongst offspring of parents of given genetic constitution. The relative proportions of different kinds of offspring resulting are the *Mendelian ratios*, whose discovery in actual breeding experiments led Mendel to formulate the basic laws of inheritance.

The result of the variety of gametes produced by an individual is variety of offspring produced by any pair of parents. The chance of two such offspring being genetically alike is very remote, and (barring polyembryony, q.v.) probably no two individuals of a sexually reproducing population are genetically alike. This means that a great range of variation is always available for evolutionary change by natural selection (q.v.). This variety depends on the genes behaving as independent units, which segregate in gamete formation. If the genetic material received by an organism from its parents really blended together, then all the gametes produced by that organism would be alike, constituted of the same blend. All the variety within a population does not of course depend upon this diversity of chromosomal genes. Some depends on environmental differences during development of individuals. A small amount may depend on genetic material carried in the cytoplasm and not distributed by the segregating mechanism of the chromosomes (see *Plasmagenes*). Differences between species are probably fundamentally of the same kind as differences within species.

GENECOLOGY. Study of genetical composition of plant populations in relation to their habitats.

GENE FREQUENCY. Frequency of occurrence of a given kind of gene in a population in relation to frequency of all its alleles.

GENERALIZED. Not specialized (q.v.).

GENERIC. (Adj.). Of genus (q.v.).

GENETIC. Concerned with the genes (q.v.); or concerned with development.

GENETICS. Study of heredity and variation; of the resemblances and differences between organisms. See *Gene.*

GENOME (GENOM). The set of all *different* chromosomes found in each nucleus of a given species. A haploid nucleus has one genome.

GENOTYPE. Genetic constitution (the particular set of alleles present in each cell of an organism), as contrasted with the characteristics manifested by the organism (phenotype, q.v.).

GENUS. One of the kinds of group used in classifying organisms. Consists of a number of similar species; occasionally of one species only. Similar genera are grouped in a family. See *Binomial Nomenclature*, *Classification*.

GEOLOGICAL PERIODS. Main fossil-bearing geological periods and their ages (it must be emphasized that the latter are only approximate) are as follows:

Era	Period	Million years since beginning of period
Cenozoic	Pleistocene	1
	Tertiary	70
Mesozoic	Cretaceous	135
	Jurassic	180
	Triassic	225
Palaeozoic	Permian	270
	Carboniferous	350
	Devonian	400
	Silurian	440
	Ordovician	500
	Cambrian	600

The Tertiary is further subdivided into the following periods (in brackets, number of million years since beginning of period): Pliocene (10), Miocene (25), Oligocene (40), Eocene (70).

Apart from a very few pre-Cambrian fossils, the fossil record of animals begins with a great variety of invertebrates in Cambrian rocks; that of plants with the occurrence of Algae in Ordovician rocks. Primitive land plants are known from the Silurian period. By late Devonian times Arthropoda, including insects, were established on land; and the first terrestrial vertebrates (Amphibia) had appeared. Members of the Pteridophyta, many of them treelike, were now abundant. They provided an extraordinarily rich flora in Carboniferous times – the age of Pteridophytes. Gymnosperms, Bryophyta, Algae, and Fungi were also present. Tremendous accumulations of the remains of Carboniferous plants, partially decayed and subjected to intense pressure by overlying deposits, formed the coal seams of today. The Mesozoic era was the age of the great dinosaurian reptiles, and Gymnosperms were the dominant plants. Angiosperms appeared in the Jurassic, became increasingly dominant towards the end of the Cretaceous as the Gymnosperms declined, and remained the dominant plant group to the present day. The ascendancy of the mammals also began with the end of the Mesozoic.

GEOPHYTES. Class of Raunkiaer's Life Forms (q.v.).

GEOTAXIS. Taxis (q.v.) in which stimulus is gravitational force.

GEOTROPISM. (Bot.). Orientation of plant parts under stimulus of gravity. Main stems, *negatively geotropic*, grow vertically upwards and if laid horizontal their growing tips turn upwards and resume a vertical position. Main roots, *positively geotropic*, grow vertically downwards and if laid horizontal their growing tips turn downwards. See also *Diageotropism* and *Plagiogeotropism*.

GEPHYREA. Class of annelids; a few genera of marine burrowing worms. Coelom spacious; segmentation partially or completely absent; chaetae present in some species.

GERM-CELLS. Gametes (q.v.).

GERMINAL EPITHELIUM. Epithelium lining seminiferous tubules of vertebrate testis, from which during sexual maturity arise the spermatogonia and the Sertoli cells that presumably nourish the developing sperm; applied also to epithelium covering coelomic surface of vertebrate ovary, which probably forms follicle cells, though it is now doubtful whether this forms oogonia throughout sexual maturity.

GERMINAL VESICLE. Nucleus of animal oocyte during the period of cytoplasmic growth. Nucleus is in prophase of the first meiotic division and is enormously enlarged, with very little if any stainable chromatin and with prominent nucleoli.

GERM-LAYER. (Zool.). One of the main layers or groups of cells which can be distinguished in an embryo during and immediately after gastrulation (q.v.). These layers are roughly similar in arrangement, and in ultimate differentiation, in most animals. There are two in diploblastic animals, endoderm (q.v.) and ectoderm (q.v.). In triploblastic animals there is in addition the mesoderm (q.v.). There is general agreement as to which layer gives rise to which tissues, e.g. nervous tissue is characteristically ectodermal, cartilage both mesodermal and ectodermal. (But it is not agreed to which germ layer the chordate notochord belongs). It is now known that the equating of the layers in different groups cannot be pushed to extremes, as it formerly was, by supposing that a given sort of tissue is in all groups of animals entirely derived from a given germ-layer. Though the general architecture is similar, there are also numerous differences between groups.

GERM-PLASM. A particular sort of protoplasm which Weismann (1834-1914) suggested was transmitted substantially unchanged from generation to generation via the germ-cells, giving rise in each individual to the body-cells (soma) but itself remaining distinct and unaffected by the environment of the individual (*theory of continuity of the germ-plasm*). Roughly equateable with modern genes (q.v.) which however are not restricted to germ cells but occur in all cells. A special cytoplasm in ovum which during embryonic development becomes restricted to future germ-cells of adult, a sort of germ-plasm which does not however give rise to the soma, has been demonstrated in some animals, though not in others.

GESTATION PERIOD. Length of time from conception to birth in viviparous animal.

GIANT FIBRE. Nerve fibre (q.v.) of very large diameter (up to 1 mm. in squids) occurring in many invertebrates (e.g. annelids, crustaceans, cephalopods), and some vertebrates, providing very rapid conduction of impulses for sudden locomotion. Can be either a single enormous cell, or the result of fusion of many cells.

GILL. (1) (Bot.). See *Lamella*, *Gill-fungi*. (2) (Zool.). Respiratory organ of aquatic animals: a projection of external surface of body or of internal layer of gut, well supplied with blood; through gill surface interchange by diffusion of oxygen and carbon dioxide between water and blood occurs. Often complex form, providing large surface area. Gills of most fishes are *internal gills* within gill slits, consisting of projections from wall of pharynx, i.e. endodermal. Gills of amphibian and dipnoan larvae, and of adults of some urodeles, are *external gills*, from epidermis of gill-slits, i.e. ectodermal.

GILL-BAR. Tissue separating adjacent gill-slits in chordates, containing blood-vessels, nerves, and skeletal support.

GILL-CLEFT. (1) Inpushing of epidermis corresponding to gill-pouch (q.v.); (2) Synonymous with gill-slit (q.v.).

GILL FUNGI. Fungi belonging to the family Agaricaceae of the Basidiomycetes; possessing a characteristic fruit-body consisting of a stalk (stipe) supporting a cap (pileus) on the under surface of which are radially arranged gills (lamellae) bearing the hymenium, e.g. mushrooms, toadstools.

GILL-POUCH. Outpushing of side-wall of pharynx towards epidermis in all chordate embryos, precursor of gill slit in fish and some Amphibia, but breaks through to exterior only temporarily or never in terrestrial vertebrates. A series of gill pouches occurs on each side, one behind the other, corresponding to segmentation of head. See *Spiracle*. Term also used for gill-slit of adult when this is unusually sac-like (cyclostomes).

GILL-SLIT. Opening from side of pharynx to exterior in aquatic chordates. Usually vertically elongated, several lying in a series one behind the other on each side. Primitively probably concerned in filtering food particles from water pumped through them by action of cilia (as in tunicates and amphioxus). In fish and some Amphibia serve for respiration, the water, which is pumped through by muscular action, oxygenating blood in the gills (q.v.). Absent in adult tetrapod vertebrates, except in a few Amphibia, but occur in many tetrapod embryos. See also *Gill-Pouch*. Presence of gill-slits, or at least of gill-pouches and corresponding epidermal grooves at some stage of development, is a characteristic of phylum Chordata.

GINKGOALES. Order of Gymnospermae that flourished during the

Mesozoic and now represented by one living member often referred to as a 'living fossil', *Ginkgo* (maidenhair tree). Leaves wedge-shaped, deciduous. Dioecious; microsporophylls arranged loosely in a catkin-like male cone; female cone consisting of a stalk bearing, usually, two ovules. Male gametes motile by cilia.

GIZZARD. Part of alimentary canal in many animals, where food is broken up. Precedes main digestive region. Very muscular walls; often contains hard 'teeth' attached to walls, e.g. Crustacea, or hard swallowed objects, as in birds.

GLAND. An organ (sometimes a single cell) whose main function is to build up one or more specific chemical compounds (secretions) which are passed to the outside of the gland.
(Bot.). Superficial, discharging secretion externally, e.g. glandular hair (lavender), nectary (q.v.), hydathode (q.v.); or embedded in tissue, occurring as isolated cells containing the secretion, or as layer of cells surrounding intercellular space (secretory cavity or canal) into which secretion is discharged, e.g. resin canal of pine.
(Zool.). Most glands are *exocrine*, i.e. their secretions are discharged, usually through a tube or *duct*, on to a surface, either the outer (epidermal) surface of the body, e.g. sweat-glands of mammals, or the inner surface of the gut, e.g. digestive glands. There are also in some animals *endocrine glands*, secreting hormones (q.v.) directly into blood.

GLENOID CAVITY. Cup-like hollow on each side of pectoral girdle (on scapula and, when present, coracoid) into which head of humerus fits, forming shoulder-joint, in tetrapod vertebrate.

GLIA (NEUROGLIA). Supporting tissue of vertebrate central nervous system, consisting mainly of cells with long fibrous processes, derived from embryonic neural tissue; and mesodermal cells (microglia) closely similar to macrophages.

GLOBIGERINA OOZE. Calcareous mud, covering huge areas of ocean-bottom (about $\frac{1}{3}$ of whole sea floor) mainly from shells of Foraminifera (Protozoa), *Globigerina* being important genus.

GLOBULINS. Group of proteins, all heat-coagulable, soluble in dilute salt solution but not in water. Widely distributed in cells of plants and animals. Amongst globulins are the main proteins of plant seeds; and antibodies.

GLOMERULUS. Of vertebrate kidney; small bunch of capillaries, covered by thin epithelium, which projects into Bowman's capsule (q.v.). Capillaries are supplied with blood by an arteriole, and in birds and mammals discharge into another (efferent arteriole) which in turn supplies capillaries of uriniferous tubule (q.v.). Much of the dissolved substances, and much of the water, of blood flowing through glomerulus filters into capsule and passes down tubule (where most is reabsorbed into bloodstream again). Plasma proteins and blood corpuscles do not filter into capsule.

GLOSSOPHARYNGEAL NERVE. Ninth cranial nerve of vertebrates.

In mammals mainly concerned with both sensory and motor aspect of the swallowing reflex, and taste-buds of back of tongue. A dorsal root (q.v.).

GLOTTIS. Opening of trachea into pharynx of vertebrate. Can usually be closed by muscles. In mammals, opening between vocal cords. See *Larynx*, *Epiglottis*, *Trachea*.

GLUCOSE (DEXTROSE). A 6-carbon-atom sugar (a hexose) widely distributed in plants and animals, particularly in compounds as disaccharides, e.g. sucrose, and polysaccharides, e.g. starch, cellulose, glycogen. Splitting of glucose, ultimately to CO_2 and water, involving intermediary combining with phosphate is a major energy-source for metabolic processes. In green plants, glucose is product of photosynthesis, from CO_2 and water; it is stored as starch. In animals, glucose is obtained mainly from digestion of di- and polysaccharides and deamination of amino acids; it is stored as glycogen. See *Blood Sugar*.

GLUME (STERILE GLUME). Chaffy bract, pair of which occur at base of grass spikelet, enclosing it.

GLYCOGEN. 'Animal starch'. A soluble polysaccharide (q.v.) built up of numerous glucose molecules. Carbohydrate is stored by animals and fungi as glycogen. In vertebrates occurs especially in liver and muscles.

GLYCOLYSIS. Conversion of glucose into lactic acid, involving a complex system of enzymes and co-enzymes.

GLYCOSIDE. Compound yielding by enzymatic or acid hydrolysis a sugar (most frequently glucose) and one or more other substances. Widely distributed in plants, e.g. anthocyanins (q.v.), amygdalin, sinigrin, digitalin, strophanthin, indican. Formation of some glycosides thought to render unwanted substances chemically inert; other functions ascribed to particular glycosides include formation of food reserves, control of invasion by pathogens. When the other substance is also a sugar, di- tri- or polysaccharides result.

GNATHOSTOMATA. Vertebrates with jaws. A grouping of vertebrate classes sometimes used in classification as a subphylum in contrast to Agnatha.

GNETALES. Order of Gymnospermae (q.v.) showing resemblances to Angiospermae in the presence of true vessels in the secondary wood and absence of archegonia in the ovule (except in *Ephedra*). Comprises three genera, *Ephedra*, *Welwitschia*, and *Gnetum*, which differ remarkably in habit and geographical distribution.

GOBLET CELL. Pear-shaped cell, present in some epithelia, the part near the free surface of the epithelium being swollen with mucin, which is secreted from time to time on to the surface.

GOLGI APPARATUS (G. BODY, G. MATERIAL). Local clump of material present in cytoplasm of most animal cells, and possibly in some plant cells. It appears in the electron microscope as a system of membrane-surrounded vacuoles. It can be blackened by staining techniques using silver or osmium, and then often appears as a network. Function unknown, but often seems associated with formation of secretions. See Fig. 1, p. 44.

GONAD. Organ of animals which produces gametes. In some animals produces hormones too. See *Ovary; Testis; Ovotestis.*

GONAD HORMONES. Sex hormone produced by gonad. In vertebrates they are steroids. See *Androgen, Oestrogen, Progesterone.*

GONADOTROPHIC (GONADOTROPIC) HORMONES (GONADOTROPHINS). Hormones secreted by anterior lobe of pituitary (q.v.) of vertebrates, and by mammalian placenta, which control activity of gonads. Changes in output of gonadotrophic hormones from the pituitary are responsible for the onset of sexual maturity, for breeding season rhythms, and for oestrous cycles (q.v.). See *Follicle-Stimulating Hormone, Luteinising Hormone.*

GRAAFIAN FOLLICLE. Fluid-filled spherical vesicle in mammalian ovary containing an oocyte (q.v.) attached to its wall. Differs from ovarian follicle of other vertebrates in presence of cavity. Cavity first appears towards end of period of cytoplasmic growth of oocyte, amongst the follicle cells which closely surround oocyte. Cavity enlarges during early part of oestrous cycle (q.v.) of those mammals which have such a cycle, separating follicle cells into an external layer and a layer adherent to oocyte. Enlargement continues until follicle bursts on to surface of ovary, discharging the oocyte (ovulation, q.v.). A considerable number of follicles which grow during the first part of an oestrous cycle never ovulate, but degenerate. Growth of follicles is under influence of pituitary gland. See *Follicle-Stimulating Hormone.* Follicle cells are probably mainly responsible for oestrogen-production of ovary. After ovulation follicle becomes corpus luteum (q.v.) in most mammals.

GRAFT. (1) To graft is to induce union between tissues normally separate. (2) A graft is a relatively small part of a plant or animal which is grafted on to a relatively large part. The graft may be moved (transplanted) from its normal position to another place on the same organism (*autoplastic* grafting, autografting); or it may be moved to a different individual of the same species (*homoplastic* grafting, homografting) or of a different species (*heteroplastic* grafting, heterografting). In plants, the part grafted on is called the *scion,* that to which it is united the *stock.* In animals these individuals are respectively called *donor,* and *host* or *recipient.* See *Chimaera, Transplantation.*

GRAFT-HYBRID. See *Chimaera.*

GRAM'S STAIN. Stain used in bacteriology. Differentiates many kinds of (*gram-positive*) bacteria which take the stain, e.g. *Staphylococcus Streptococcus,* from others (*gram-negative*) which do not, e.g. *Gonococcus,* typhoid bacteria. Staining is due to presence in bacteria of a nucleo-protein.

GRANULOCYTE. Polymorph (q.v.).

GRAPTOLITES. Fossilized skeletons of extinct colonial animals found in Cambrian, Ordovician, and Silurian rocks. Relationships of the group are obscure; some zoologists regard them as Coelenterata; some as related to Hemichordata.

GREY MATTER. Tissue of vertebrate central nervous system containing numerous cell-bodies and dendrites of nerve cells; and terminations of nerve-fibres forming synapses with these; together with glia and blood-vessels. Occurs mainly as an inner layer surrounding central canal; but occurs also as a superficial layer (cortex) on cerebellum and cerebral hemispheres in some vertebrates. It is the tissue of the nuclei or centres. The co-ordinating work of the C.N.S. is done in grey matter by the large numbers of synapses present. Cf. *White matter.*

GROWTH. Increase in size. In botany usually includes also differentiation of cells, such as accompanies, or follows, increase in plant size, even though such differentiation contributes nothing to size change.

GROWTH, GRAND PERIOD OF. (Bot.). Expression applied to the total period of enlargement of a cell, an organ or part of an organ of a plant. During this period growth is slow at first, gradually increasing to a maximum and then falling off to zero as maturity is reached.

GUARD CELLS. (Bot.). Specialized, crescent-shaped epidermal cells occurring in pairs surrounding stomata. See *Stoma.*

GUT. Alimentary canal (q.v.).

GUTTATION. Excretion of drops of water by plants through hydathodes (q.v.); occurring usually under conditions of high humidity.

GYMNOPHIONA. Apoda (q.v.).

GYMNOSPERMAE. Conifers and their allies; sub-division of Spermatophyta. Primitive seed plants with many fossil representatives. Distinguished from the other group, Angiospermae, by having the ovules borne unprotected on surface of megasporophylls. Micro- and megasporophylls usually arranged in cones. Gametophyte generations reduced. Female gametophyte multicellular (female prothallus), with few exceptions, bearing archegonia. Male gametophyte initiated by pollen grain (microspore), consisting of varying number of prothallial cells and pollen tube containing male gametes; in some orders these are motile, in others non-motile. Characteristically without vessels in xylem (except in order Gnetales). See *Coniferales, Cycadales, Ginkgoales, Gnetales, Bennettitales, Cordaitales, Cycadofilicales.*

GYNANDROMORPH. Abnormal organism. Organism of mixed sex; some of tissues are genetically and structurally female, others genetically and structurally male. Can occur, e.g. through disturbance of genetic sex-determination mechanism during embryonic cleavage, as by loss of an X-chromosome (q.v.) in one blastomere and consequently in all its descendants. One half of such an organism can be male, one half female. Well-known in insects; occurs also in birds and mammals. Cf. *Intersex*.

GYNOBASIC (of a style), arising from base of ovary (due to infolding of ovary wall during development), e.g. white dead nettle.

GYNODIOECIOUS. Having female and hermaphrodite flowers on separate plants.

GYNOECIUM (PISTIL). Collective name for the carpels of a flower.

GYNOMONOECIOUS. Having female and hermaphrodite flowers on the same plant.

H

HABITAT. Place with particular kind of environment inhabited by organism(s), e.g. the sea shore.

HAEMOCOEL. Body cavity (q.v.) which is really expanded part of blood-system, containing blood. Well developed in Arthropoda and Mollusca where coelom is small. Unlike coelom it never communicates with exterior, and never contains germ cells (cf. *Coelom*).

HAEMOCYANIN. Blue respiratory pigment (q.v.) present in blood of some molluscs and arthropods. It is a protein containing copper.

HAEMOGLOBIN. Respiratory pigment (q.v.) occurring in red blood cells of vertebrates and sporadically elsewhere among animals, e.g. in blood of earthworms, muscles of some roundworms. Occurs also in root-nodules of nitrogen-fixing leguminous plants. Scarlet in oxygenated form, bluish-red deoxygenated. Many haemoglobins are known, differing in molecular weight, absorption spectrum, oxygen-combining properties, etc. Each animal species has a different haemoglobin; and (in mammals) that of the foetus is different from that of the adult. A protein with iron in the prosthetic group; closely related to chlorophyll and to cytochrome. See *Myoglobin*.

HAEMOLYSIS. Escape of haemoglobin from red blood corpuscles owing to damage to surface membrane.

HAEMOPHILIA. Human disease in which blood-clotting is defective. Known only in males. Transmitted from mother to son. Determined by sex-linked recessive gene. Women homozygous for this gene are unknown.

HAEMOPOIESIS (HAEMATOPOIESIS). Formation of blood corpuscles. Occurs in lymphoid tissue and bone-marrow of vertebrates.

HAIR. (1) (Bot.). (*Trichome*). Single-celled or many-celled outgrowth from an epidermal cell. Functions various, e.g. absorbing (see *Root-hair*); secretory (see *Gland*); lowering transpiration rate. (2) (Zool.). Characteristic of mammal. Each hair consists of numerous cornified epidermal cells; grows at its base in hair-follicle where active cell-division occurs. See *Seta*.

HAIR-FOLLICLE. Deep pit of mammalian epidermis surrounding root of hair which projects far down into dermis. Receives duct of sebaceous gland. Muscle which erects hair is attached to it.

HALLUX. 'Big toe' of pendactyl hind-limb. On anterior or inner side (tibial side). Often shorter than other digits.

HALOPHYTE. Plant that tolerates very salty soil, a condition typical of shores of river estuaries.

HALTERE. Modified hind-wing of dipteran fly, consisting of a small projection from the body. Sensory function, probably concerned with maintaining equilibrium during flight.

HAPLOCHLAMYDEOUS. See *Monochlamydeous*.

HAPLOID. Having a single set of unpaired chromosomes in each nucleus. Characteristic of gametes; of some Sporozoa; of somatic nuclei of parthenogenetically produced males of some animals, e.g. bees; and of gametophytes and many spores of Algae, Fungi, Bryophyta, and vascular plants. See *Diploid*, *Alternation of Generations*, *Meiosis*.

HAPLONT. Haploid stage of an organism, ending with fertilization. Cf. *Diplont*.

HAPTOTROPISM (THIGMOTROPISM). (Bot.). Tropism (q.v.) in which stimulus is a localized contact, e.g. a tendril in contact on one side with a solid object, a twig for instance, curves in that direction and coils round it.

HAUSTORIUM. (1) Specialized branch of hypha of parasitic fungus that penetrates a living cell of the host and absorbs food material from it. (2) Specialized organ of parasitic plant, e.g. dodder, which penetrates into and withdraws food material from tissues of host plant.

HAVERSIAN CANALS. Channels (roughly 50μ diameter), carrying blood-vessels and nerves, which ramify throughout bone, communicating with its surface and marrow. Sheets (lamellae) of bone, and bone cells, are arranged concentrically around the canals.

HEARTWOOD. Central mass of xylem tissue in trunks of trees, with no living cells (such as xylem parenchyma and medullary ray) and no longer functioning in water conduction but only for mechanical support; with its elements frequently blocked by tyloses (q.v.) and often dark-coloured, e.g. oak, ebony, impregnated with various substances (tannins, resins, etc.) that render it more resistant to decay than surrounding sapwood.

HELIOTROPISM. Phototropism (q.v.).

HELMINTH. A worm, usually a parasitic one.

HELMINTHOLOGY. Study of parasitic worms.

HELOPHYTES. Class of Raunkiaer's Life Forms (q.v.).

HEMICELLULOSES. Group of carbohydrates related to cellulose forming part of the cell wall of plants, especially in lignified tissue; may function as reserve food materials especially in seeds, e.g. endosperm of date seeds.

HEMICHORDATA (ENTEROPNEUSTA). Small group of marine animals, included in the Protochordata; contains worm-like burrowing *Balanoglossus*, and sedentary ciliary feeders *Cephalodiscus* and *Rhabdopleura*. They have gill slits; a structure sometimes regarded as notochord in anterior region only; nerve cord solid and superficial except in anterior region where it is hollow. Embryology suggests relationship with Echinodermata.

HEMICRYPTOPHYTES. Class of Raunkiaer's Life Forms (q.v.).

HEMIMETABOLA. Exopterygota (q.v.).

HEMIPTERA (RHYNCHOTA). Bugs. Large order of exopterygote insects, including scale-insects, leaf-hoppers, cochineal-insects, lac-insects, bed-bugs, green-flies. Usually two pairs of wings, anterior harder than the posterior. Parasites of plants and animals (whose juices they suck with their piercing mouth parts) and as such are of enormous economic importance, e.g. *Aphis* (q.v.).

HEPARIN. Substance which prevents blood clotting (q.v.) by stopping conversion of prothrombin to thrombin and by neutralizing thrombin. Extractable from various tissues, especially lung. Probably contained in mast cells, which occur in connective tissue. Sulphur-containing complex polysaccharide.

HEPATIC. (Adj.). Of the liver.

HEPATIC PORTAL SYSTEM. System of veins carrying blood from capillaries of intestine to liver in vertebrate (and amphioxus). Absorbed products of digestion (except fats, see *Lacteals*) thus go straight to liver cells which transform them chemically in various ways. See *Deamination.*

HEPATICAE. Liverworts, class of green plants grouped with Musci (mosses) in division Bryophyta. Living in wet conditions, on soil or as epiphytes (q.v.), or in water. Consisting of a thin, prostrate plant body or a creeping central axis up to a few inches long, provided with leaf-like expansions; attached to soil by rhizoids growing from under surface. Sexual organs, antheridia and archegonia, variously grouped, male gametes motile by flagella. Fertilization is followed by development of a capsule containing spores which, being shed, germinate in most forms to form a short, thalloid *protonema* from which new liverwort plants arise. Differ from mosses in the simpler organization of plant body and sporophyte.

HERB. Plant with no persistent parts above ground, as distinct from shrubs and trees.

HERBACEOUS. Having the characters of a herb.

HERBARIUM. Collection of preserved plant specimens for identification and reference purposes.

HERBIVORE. Plant-eater. Cf. *Carnivore, Omnivore.*

HERMAPHRODITE (BISEXUAL). (1) (Of a flowering plant or flower) having both stamens and carpels in the same flower. Cf. *Unisexual.* (2) (Of an individual animal) producing both male and female gametes. Among unisexual (q.v.) animals hermaphrodites may occur as aberrations.

HETEROCHLAMYDEOUS. (Of flowers), having two kinds of perianth segments (sepals and petals) in distinct whorls. Cf. *Homochlamydeous.*

HETEROCHROMATIN. Chromosome material showing its maximum staining in the resting nucleus or in prophase (q.v.) and often less dense staining during metaphase (q.v.). Contains few or no genes of major effect. Sex chromosomes, especially Y-chromosome of many animals, tend particularly to have large heterochromatic regions. Cf. *Euchromatin.*

HETERODONT. Having teeth of different kinds, as in most mammals, which have incisors, canines, molars. Cf. *Homodont.*

HETEROECIOUS. Of rust Fungi (order Uredinales of Basidiomycetes), having certain spore forms of life-cycle on one host species and others on an unrelated host species. Cf. *Autoecious.*

HETEROGAMETIC SEX. The sex, individuals of which have within each of their nuclei a pair of dissimilar sex-chromosomes (q.v.) (one X- and one Y-chromosome); or an unpaired sex-chromosome (a single X-chromosome). Heterogametic sex is usually male, but is female in Lepidoptera, birds, reptiles, some amphibia and fish, and a few plants. Cf. *Homogametic sex.*

HETEROGAMY. (1) Synonym of anisogamy (q.v.); (2) Occasionally synonym of oogamy (q.v.); (3) (Zool.). Alternation of two forms of sexual reproduction in successive generations, e.g. of parthenogenesis and syngamy as in some aphids; or of hermaphrodite and dioecious, as in some nematodes.

HETEROGRAFT. Graft originating from an animal of a different species from that which receives it. Cf. *Autograft, Homograft.*

HETEROKARYON, HETEROKARYOSIS. Association in somatic cells of nuclei of unlike genetical constitution; found in many fungi, especially in Fungi Imperfecti. Variation in relative proportions of different nuclei on different substrates provides a system of somatic variation and adaptation. Cf. *Homokaryon.*

HETEROMETABOLA. Exopterygota (q.v.).

HETEROMORPHIC. Used of alternation of generations (q.v.), particularly in Algae, meaning generations vegetatively dissimilar. Cf. *Isomorphic.*

HETEROSIS (HYBRID VIGOUR). Increased vigour of growth, fertility, etc., in a cross between two genetically different lines, as compared with growth, etc. in either of the parental lines; associated with increased heterozygosity.

HETEROSPOROUS. Having two kinds of spores, microspores and megaspores, that give rise, respectively, to distinct male and female gametophyte generations; a condition found in certain Pteridophyta (where the term was first applied) and in Spermatophyta. In the former group there is a conspicuous difference in the size of the two kinds of spores, indicated in the terms applied to them, but this is not usually found in the Spermatophyta. Here the terms indicate homologies with the small and large spores of heterosporous pteridophytes rather than differences in size. Cf. *Homosporous.*

HETEROSTYLY. Condition in which length of style differs in flowers of different plants of a species, e.g. pin-eyed (long style) and thrum-eyed (short style) primroses. Anthers in one kind of flower are at same level as stigmas of other kind. Device for ensuring cross-pollination by visiting insects. Cf. *Homostyly.*

HETEROTHALLISM. (Of Fungi, Algae) condition in which sexual reproduction occurs only through participation of two thalli, each of which is self-sterile. In Fungi, includes *morphological heterothallism* where sexes are segregated, some thalli being male, others female; and *physiological heterothallism* where interacting thalli, often labelled plus and minus strains, show no morphological difference such as might be recognized as a difference in sex. Includes forms in which both thalli bear male and female sex organs and others which have no sex organs, union between strains depending on hyphal fusions. Cf. *Homothallism.*

HETEROTRICHOUS. Type of thallus found in many filamentous algae; consisting of a prostrate creeping system from which project branched filaments.

HETEROTROPHIC. (Of an organism) requiring a supply of organic material (food) from its environment. All animals and fungi, most bacteria and a few flowering plants are strongly heterotrophic, requiring organic food substances from which to make most of their own organic constituents. As herbivores, carnivores, saprophytes, or parasites they obtain this food from other organisms, living or dead. Ultimately almost all their organic material can be traced back to the synthetic activity of *autotrophic* organisms. Heterotrophic organisms commonly obtain their energy requirements also from organic food, i.e. they are chemotrophic (q.v.); but a few (particularly flagellates) are phototrophic (q.v.).

HETEROZYGOUS. Having two different allelomorphs in the two corresponding loci (q.v.) of a pair of chromosomes (q.v.). Cf. *Homozygous.* A heterozygous organism is a *heterozygote* for the locus in question; and with respect to that locus it produces two different kinds of gametes. Its phenotype is frequently identical with that of one of the two allelomorphs in the homozygous state (see *Dominant*); but it may be more or less intermediate between those of the two homozygotes.

HEXACANTH. Onchosphere (q.v.).

HEXAPODA. Insecta (q.v.).

HEXOSE. Sugar (monosaccharide) with six carbon atoms, e.g. glucose, fructose, galactose. Combinations of hexoses make up most of biologically important disaccharides and polysaccharides.

HIBERNATION. Dormancy during winter. Occurs in many mammals, most reptiles and Amphibia, and many vertebrates, of temperate and arctic regions. Metabolism is greatly slowed, and in mammals temperature drops to that of surroundings. Some, e.g. bats, wake to feed from time to time. See *Diapause*, *Aestivation*.

HILUM. Scar on seed coat marking former point of attachment of seed to funicle.

HIND-BRAIN. Hindmost of the three divisions marked out by constrictions in the embryonic vertebrate brain. Becomes during development medulla (q.v.) and cerebellum (q.v.).

HIP-GIRDLE (PELVIC GIRDLE). Skeletal support, situated in body-wall, for attachment of hind fins or limbs of Vertebrata. In fish consists fundamentally of a curved bar of bone or cartilage, situated transversely to long axis of body, one on each side in ventral region; the two bars usually fusing mid-ventrally to form a half-loop. At middle of the length of each bar is joint with the fin. Region dorsal to joint is *ilium*, region ventral is *ischio-pubis*. In tetrapods, except the earliest Amphibia, ischio-pubis forms two bones, an anterior pubis and posterior ischium; and the ilium, which extends much further dorsally than in fish, unites with one or more sacral vertebrae (q.v.) (compare shoulder girdle), forming a complete girdle round body in this region and giving rigid support to hind-limbs for locomotion (unnecessary in fish, which are propelled by tail). See *Acetabulum*, *Innominate Bone*. Fig. 10, p. 216.

HIRUDINEA. Leeches, a class of annelids. Marine, freshwater or terrestrial; suckers at anterior and posterior ends; segmentation less clear than in chaetopods, as each segment is superficially subdivided into narrow rings; no chaetae; coelom partially obliterated; hermaphrodite; clitellum; fertilization internal; eggs develop in a cocoon. Some are blood-suckers.

HISTAMINE. Organic base, released from tissues when they are injured, causing dilation of local blood-vessels.

HISTIOCYTE. See *Macrophage*.

HISTOCHEMISTRY. Study of distribution of particular chemical substances, by specific staining methods, etc., within sections or whole mounts of tissues.

HISTOGENESIS. Differentiation of tissue.

HISTOGENS. Distinct tissue zones recognizable in apical meristems (q.v.) of many plants, especially in roots.

HISTOLOGY. Study of tissues.

HISTOLYSIS. Dissolution of tissue.

HOLOBLASTIC. (Of animal zygote) the whole undergoing cleavage (q.v.). Occurs in eggs with little or moderate amounts of yolk. Cf. *Meroblastic*.

HOLOCENE (RECENT). Geological period consisting of recent times since end of the last ice-age (about 10,000 years ago).

HOLOCEPHALI. A group of fish, including *Chimaera*. A sub-class or order of Chondrichthyes. Differ from other Chondrichthyes (see *Selachii*) in their large flat crushing teeth, and the fusion of upper jaw (palatoquadrate) to skull, both associated with diet of molluscs; and in having an operculum covering gill slits. Rare fish now, but with many fossil relatives. See *Elasmobranchii*.

HOLOGAMETE. See *Merogamete*.

HOLOMETABOLA. Endopterygota (q.v.).

HOLOPHYTIC. Feeding like a green plant, i.e. by synthesizing organic compounds from inorganic components, using the energy of sunlight by means of chlorophyll. Cf. *Holozoic*. See *Autotrophic*.

HOLOTHUROIDEA. Sea-cucumbers; class of Echinodermata. Body cylindrical, soft.

HOLOTYPE. Type specimen (q.v.).

HOLOZOIC. Feeding in an animal-like manner, i.e. by eating other organisms or solid organic material elaborated by them. Cf. *Holophytic*. See *Heterotrophic*.

HOMEOSTASIS. Maintenance of constancy of internal environment (q.v.).

HOMINID. Man (*Homo*) and man-like fossils, e.g. *Pithecanthropus*. These together make up the family Hominidae (order Primates).

HOMO. Genus of catarrhine primates whose only living representative is man (*Homo sapiens*); but it contains several extinct species, e.g. Neanderthal man. *H. sapiens* is distinguished anatomically from living catarrhines by large brain (about 1500 c.c.); absence of brow ridges; chin prominence; teeth in each jaw arranged in a smooth curve, with small canines; foot very different from hand, with reduced toes, big-toe not opposable to others. Distinguished in behaviour by walking upright or almost so, by using tools with hands, by speech, and by the cultural tradition derived therefrom. Extinct species possess some of these characters in nearly human degree, while other characters are ape-like. See *Pithecanthropus*.

HOMOCHLAMYDEOUS. (Of flowers), having perianth segments of one kind (tepals) in two whorls. Cf. *Heterochlamydeous*.

HOMODONT. Having a set of teeth all of the same kind. As in most vertebrates other than mammals. Cf. *Heterodont*.

HOMOGAMETIC SEX. The sex, individuals of which have within each of their nuclei a pair of similar sex-chromosomes (q.v.) (X-chromosomes). Cf. *Heterogametic sex*.

HOMOGAMY. Condition in which male and female parts of a flower mature simultaneously. Cf. *Dichogamy*.

HOMOGRAFT. Graft originating from an animal of the same species as the recipient, but not from the recipient itself. Cf. *Autograft*, *Heterograft*.

HOMOIOTHERMIC. 'Warm-blooded'. Maintaining a constant body temperature, raised above that of usual surroundings. Characteristic of birds and mammals. Cf. *Poikilothermic*.

HOMOKARYON. (Of fungus hypha or mycelium) composed of cells whose nuclei are of identical genetical constitution. Cf. *Heterokaryon*.

HOMOLOGOUS. An organ of one animal is said to be homologous with an organ of another when both have a fundamental similarity of structure and/or position relative to other organs, manifested especially during embryonic development; regardless of their functions in the adult, which may be very different, e.g. mammalian ear-ossicles (q.v.) are homologous with bones concerned in fish jaw attachment. The degree of similarity which constitutes 'fundamental similarity' is a matter of opinion, so that homology is not a clear-cut concept. The similarity is assumed to be due to descent of the organisms from a common ancestor. The presumption is that a developmental process occurring in a common ancestor has become modified in two different directions in two descendant species, without the divergence obscuring the common origin. Evolution has evidently generally occurred, not by sudden origin of quite new embryonic processes, but by slight modification of existing ones, the changes taking effect particularly at later stages of development; resulting in the persistence of the similarities between related organisms, especially between their embryos, denoted by homology. Homology of organs implies relationship of the organisms bearing the organs, and it is the main concept of evolutionary comparative anatomy. Cf. *Analogous*.

HOMOLOGOUS CHROMOSOMES. Chromosomes which contain identical sets of loci (q.v.) ('homologous' is applied also to parts of chromosomes). The two chromosomes of each of the pairs normally found in all somatic cells of animals and of plant sporophytes are homologous. Two homologous chromosomes or parts of chromosomes have a strong attraction for each other during early stages of meiosis (q.v.) and undergo pairing.

HOMOSPOROUS. Having one kind of spore that gives rise to gametophyte generation bearing both male and female reproductive organs. Cf. *Heterosporous*. Characteristic of certain members of *Pteridophyta*. Exceptionally, as in Equisetales (q.v.), gametophyte plants may be dioecious in certain environmental conditions.

HOMOSTYLY. Condition (the common one) in which flowers of a species have styles of one length as opposed to *Heterostyly* (q.v.).

HOMOTHALLISM. (Of Fungi) condition in which sexual reproduction occurs in a colony derived from a single spore, i.e. each thallus is self-fertile. Cf. *Heterothallism*.

HOMOZYGOUS. Having identical genes (i.e. not having different allelomorphs) in the two corresponding loci of a pair of chromosomes (q.v.). Cf. *Heterozygous*. A homozygous organism is a *homozygote* for the loci in question. It has a phenotype characteristic of the particular allelomorph (but see *Penetrance*), whether or not the allelomorph is dominant or recessive to other allelomorphs.

HOOKWORMS. Nematode parasites of man (*Necator americanus* and *Ancylostoma duodenale*). Of very great medical and social importance. Common in parts of Africa, India, China, East Indies, South America, Southern parts of North America. About 1 in. long; live in intestine; eggs, liberated in faeces, giving rise in suitable conditions to active larvae which penetrate human skin, chiefly of feet. Heavy infections cause anaemia and physical and mental retardation.

HORMONE. Organic substance produced in minute quantity in one part of an organism and transported to other parts where it exerts a profound effect.

Hormones play a prominent role in *plant* growth, influencing, e.g. cell elongation, root production, movement of organs as in geotropic and phototropic curvature, dominance of certain parts of plants over others, activation of cambium, flower and fruit production. They are called auxins, growth hormones, phytohormones, or growth-promoting substances. Method of transport within the plant not yet understood. Substances with hormone activity are obtained from many plant and animal sources, e.g. tips of coleoptiles in grain seedlings, leaves, shoots, root tips, pollen grains, fruits, seeds, Fungi, Bacteria, human saliva, and lastly the richest source, human urine. In addition to these naturally occurring substances, a large number of synthetic organic compounds produce similar responses. Hormones in sufficient concentration can exert a toxic action and different species of plants are affected to different degrees by the same hormone. This differential effect is now being used as a method of weed control, e.g. in cereal crops and on lawns.

In *animals* hormones are usually secreted by endocrine glands directly from gland cells into blood stream. Various hormones of different composition and activities are produced in different glands, e.g. see *Adrenal, Insulin, Pituitary, Thyroid, Secretin.* Occur conspicuously in vertebrates, but known in some invertebrates, e.g. see *Corpora allata.*

HORSETAIL. See *Equisetales.*

HOST. (1) Organism infected by a parasite. *Definitive host*: that in which an animal parasite attains sexual maturity. *Intermediate host*: one which is essential for the life cycle of an animal parasite, but in which it does not become sexually mature. (2) Animal (especially embryo) into which a graft (q.v.) is experimentally transplanted.

HUMERUS. Bone of proximal part of tetrapod fore-limb (bone of upper arm of man). See Fig. 7, p. 175.

HUMUS. Complex organic matter resulting from decomposition of plant and animal tissue in the soil, which gives to surface layer of soil its characteristically dark colour; of great importance for plant growth. Colloidal, improving texture and waterholding capacity of soil and forming a reservoir of mineral nutrients which are absorbed by humus and prevented from being leached away.

HYBRID. Plant or animal resulting from a cross between parents that are genetically unlike; often restricted to the offspring of two different species or of well-marked varieties within a species.

HYBRID SWARM. Continuous series of forms resulting from hybridization of two species followed by crossing and backcrossing of subsequent generation. Often very variable owing to segregation.

HYBRID VIGOUR. Heterosis (q.v.).

HYDATHODE. Water-excreting gland occurring on the edges or tips of leaves of many plants.

HYDATID CYST. A cyst formed by larva (cysticercus stage) of certain tapeworms in, e.g. sheep, man. Consists of a fluid-filled sac which may be large enough to hold a gallon. Contains innumerable larval heads each of which if swallowed by a dog may produce a tapeworm.

HYDRA. Genus of small freshwater hydrozoan coelenterates; ubiquitous, few species. No alternation of generations (q.v.).

HYDRANTH. See *Polyp*.

HYDROID. Member of one of the orders of Hydrozoa (q.v.). Colonial animals forming tuft-like growths on seaweeds, etc.

HYDROPHYTE. (1) Plant whose habitat is water or very wet places; characteristically possessing aerenchyma (q.v.). Cf. *Mesophyte*, *Xerophyte*. (2) Class of Raunkiaer's Life Forms (q.v.).

HYDROPONICS. System of large-scale plant cultivation developed from 'water-culture' methods of growing plants in the laboratory, in which their roots dip into a solution of nutrient salts or in which the plants are allowed to root in some relatively inert material, e.g. quartz sand, which is irrigated with the nutrient solution.

HYDROSERE. Type of sere (q.v.).

HYDROTROPISM. Tropism (q.v.) in which the stimulus is water.

HYDROZOA. Hydroids, stinging corals, Portuguese-man-of-war, etc. Class of Coelenterata (of subphylum Cnidaria). Individual polyps and medusae usually quite small, though fairly bulky colonies occur; usually well-marked alternation of generations; coelenteron simple; gonads ectodermal. Cf. *Actinozoa*, *Scyphozoa*.

HYMENIUM. Layer of regularly arranged spore-producing structures found in fruit bodies of many ascomycete and basidiomycete Fungi.

HYMENOPTERA. Order of endopterygote insects, including bees, wasps, ants. Two pairs of wings, membranous, coupled together; larva generally grub-like. Some members very highly specialized for social life; some others are, as larvae, parasites of insects, e.g. ichneumon flies.

HYOID ARCH. Visceral arch next behind jaws of vertebrates. Dorsal part forms hyomandibula (q.v.); ventral part in adult tetrapods forms *hyoid bone*, usually supporting tongue. Contains facial nerve; and forms many muscles of the face. See *Placodermi*, *Spiracle*.

HYOMANDIBULA. Dorsal element (bone or cartilage) of hyoid arch which takes part in jaw attachment in most fish (see *Hyostylic Jaw-*

Suspension), and in tetrapods becomes columella auris (stapes).

HYOSTYLIC JAW-SUSPENSION. Found in most fish. Upper and lower jaws on each side are attached at their hinge to one end of the hyomandibula, the other end of which is anchored to the neurocranium. See *Autostylic-Jaw Suspension.*

HYPERPARASITE. Organism which lives parasitically in or on another parasite.

HYPERPLASIA. Increase in amount of tissue by increase in number of cells, which individually keep their usual size (cf. *Hypertrophy*), e.g. when part of mammalian liver is removed, rest undergoes hyperplasia. In plants, occurs in response to certain parasites.

HYPERTONIC. (Of a solution), having a concentration such that it gains water by osmosis (q.v.) across a membrane from some other specified solution. When the membrane and the other solution are unspecified, they are taken to be the plasma membranes and the interior of cells respectively. Cf. *Hypotonic*, *Isotonic.*

HYPERTROPHY. (1) Increase in size of tissue or organ by increase in size of individual elements (cells or collagen fibres) without increase of their numbers (cf. *Hyperplasia*), e.g. in exercised muscle; uterine muscle during later part of pregnancy; in plants, as response to certain parasites. (2) Sometimes, increase in size of tissue or organ, mechanism unspecified, but often hyperplasia. See *Compensatory Hypertrophy.*

HYPHA. Filament of a fungus thallus, composed of one or more cylindrical cells; increases in length by growth at its tip and gives rise to new hyphae by lateral branching.

HYPOCOTYL. Part of seedling stem below cotyledons.

HYPODERMIS. Layer of cells immediately beneath epidermis of leaves of certain plants, often mechanically strengthened, e.g. pine, forming an extra protective layer; or forming water storage tissue.

HYPOGEAL. (Of cotyledons) remaining underground when the seed germinates, e.g. broad bean, pea, Cf. *Epigeal.*

HYPOGLOSSAL NERVE. Nerve of vertebrate supplying muscles below pharynx, or in amniotes, tongue. A ventral root (q.v.). Twelfth cranial nerve of amniote vertebrates and fossil amphibians, but arises from spinal cord in living anamniotes.

HYPOGYNOUS. See *Receptacle.*

HYPONASTY. (Bot.). More rapid growth of lower side of an organ, e.g. in a leaf, resulting in upward curling of leaf-blade. Cf. *Epinasty.*

HYPOPHYSECTOMY. Surgical removal of pituitary body.

HYPOPHYSIS. (1) (*Hypophyseal Pouch*) Median ectodermal inpushing of embryonic vertebrate head, just in front of buccal membrane, which fuses with infundibulum, forming pituitary body (q.v.). (2) Synonymous with pituitary body.

HYPOTHALAMUS. Floor and sides of brain of vertebrates just behind attachment of cerebral hemispheres. Derived from fore-brain. In mammals known to contain centres co-ordinating, amongst other things, manifestations of rage and mechanism of body temperature control, both of which predominantly involve sympathetic nervous system. Pituitary is immediately below hypothalamus; latter controls secretion of anti-diuretic hormone and oxytocin, probably supplying them to pituitary.

HYPOTONIC. (Of a solution) having a concentration such that it loses water by osmosis (q.v.) across a membrane to some specified other solution. When the membrane and the other solution are unspecified they are taken to be the plasma membranes and interior of cells respectively. Cf. *Hypertonic*, *Isotonic*.

I

ICHTHYOSAURIA. A fossil order of Reptilia. Lived during Mesozoic. Marine, fish-like in general appearance and mode of swimming, with paddle-like limbs, dorsal fin, large vertical fish-like tail-fin with spinal column extended into lower lobe. Up to forty feet long. Probably viviparous, reproducing entirely in the water, which they doubtless never left.

IDENTICAL TWINS. Monozygotic twins (q.v.).

ILEUM. That part of small intestine of mammals preceding the large intestine.

ILIUM. Dorsal part of hip-girdle, which in tetrapods is jointed to one or more sacral vertebrae (q.v.) so giving stability to attachment of hind-limbs. See Fig. 10, p. 216.

IMAGINAL DISC. Thickening of epidermis, together with underlying mesenchyme, in pupa of holometabolous insect which at metamorphosis gives rise to an adult organ, the larval structure being destroyed.

IMAGO. Adult sexually mature insect.

IMBRICATE. (Of leaves, petals, etc.), overlapping.

IMMUNITY. Ability of an animal or plant to resist infection by parasitic organisms. An essential requirement for survival, since most animals and plants are perpetually menaced by viruses, bacteria, fungi, and parasitic animals. Immunity of animals is due to many different mechanisms, such as impervious skin, antiseptic stomach (due to acid), activity of phagocytes (q.v.), and chemical defence by antibodies (q.v.). Since great attention has been given to defence by antibodies, which can be artificially induced to appear in the body, the term immunity, and the subject-matter of the science of immunology, is often restricted to induced immunity. Immunity in plants is (*a*) due to structural features, e.g. waxy surface preventing wetting and consequently development of pathogens, thick cuticle preventing entry of germ-tubes of fungus spores; or is (*b*) protoplasmic, the protoplast being an unfavourable medium for further development of a pathogen; or is (*c*) acquired immunity, used with respect to virus diseases and applied to (1) recovery from an acute disease, and (2) resistance conferred against virulent strains by presence of avirulent ones. It is a non-sterile type of immunity which depends on persistence of active virus in the recovered or protected plants. Freedom from a second attack of an acute disease, or protection from the effects of virulent strains, persists only as long as the plants are infected. Plants are not known to produce antibodies.

IMPLANT. (Zool.). Material, whether tissue or not, artificially placed in a embryo (an operation employed to discover an induction or evocation) or in an adult.

IMPLANTATION (NIDATION). Attachment of mammalian embryo (blastocyst, q.v.) to lining of uterus, preparatory to forming placenta. In primates embryo dissolves lining epithelium of uterus, and embeds itself in substance of uterine wall.

IMPULSE. The 'message' which is conducted along a nerve-fibre. It is fundamentally the same in all nerve-fibres. It is a travelling wave of chemical and physical events involving particularly the surface membrane of the fibre. It moves at between 1 and 100 metres per second (speed depending on species of animal and on the diameter of, and amount of myelin around, the nerve fibre; and on temperature and other conditions at the time). The energy for the impulse is provided locally along the course of the nerve fibre, and not by the stimulus which sets the impulse going. Consequently the characteristics of the impulse (action potential (q.v.), speed, etc.) are unaffected by the nature or intensity of the stimulus. See *All-or-None Law*. The impulse runs without loss of vigour; and wherever in the nerve cell it starts, it travels right through all the branches of that nerve-cell (see *Nerve-Net*). Each impulse occupies up to an inch or two of the length of the nerve-fibre at any instant. When it has passed a given place in the nerve, that place is, for a few thousandths of a second, refractory, i.e. it will not transmit another impulse. A succession of impulses in one nerve fibre must therefore be spaced out, never being closer than the refractory zone. Most stimuli, including stimulation via a receptor, set off a train of successive impulses in a nerve-cell, rather than a single impulse; and different stimuli can therefore produce different conduction in a nerve-fibre by affecting not the individual impulse, but the number and frequency of successive impulses in a train of impulses.

INBREEDING. Reproduction by the mating of closely related individuals as opposed to *Outbreeding* by the mating of less related individuals.

INCISOR. Chisel-shaped tooth in front of mouth of mammals. Primitively three on each side of upper and lower jaws. Gnawing teeth of rodents which continually grow in length and tusks of elephants are modified incisors.

INCOMPATIBILITY. (1) In flowering plants, failure to set seed (i.e. of fertilization and subsequent development of embryo) after pollination (self or cross) has taken place. Due to inability of pollen tubes to grow down style. In physiologically heterothallic (q.v.) fungi, failure to reproduce sexually in single or paired cultures. In both cases the incompatibility is genetically determined – it is due to the possession by the individual plants or thalli of genes which prevent the fusion of likes; and the result is the same, namely, to promote outbreeding (q.v.). (2) Also used in horticulture to mean inability of scion to make a successful union with stock.

INCUS. Mammalian ear-ossicle (q.v.) representing quadrate of other vertebrates.

INDEHISCENT. (Of fruits), not opening spontaneously to liberate their seeds, e.g. hazel-nut.

INDEPENDENT ASSORTMENT. The chance distribution to the gametes of alleles, distribution of members of one pair having no influence on distribution of members of another. So that if an individual has one pair of alleles, A and a, and another pair, B and b, it will form approximately equal numbers of gametes of the four possible chance combinations of one member from each pair; pair: A and B, A and b, a and B, a and b. Mendel's Second Law asserts this, but it requires the qualification that independent assortment does not apply to linked genes (see *Linkage*), but only to those lying on different chromosomes, since it is due to the independent behaviour of chromosome pairs during meiosis (q.v.).

INDIGENOUS. (Of organisms) native to a particular area, not introduced.

INDUCTION. Influence of one embryonic tissue (or of an artificial implant) on another embryonic tissue, directing its differentiation. A tissue whose differentiation is induced by another shows dependent differentiation (q.v.). See *Evocation.*

INDUSIUM. Membranous outgrowth from underside of leaf which in certain ferns covers and protects a group (sorus) of developing sporangia.

INFERIOR OVARY. See *Receptacle.*

INFLAMMATION. The local response to any local injury in a vertebrate. Consists of (i) dilatation of blood-vessels; (ii) invasion by leucocytes; (iii) passage of blood proteins and fluid through capillary walls into tissue spaces. Functions probably as protective mechanism (by means of leucocytes and antibodies) against any invading bacteria.

INFLORESCENCE. Flowering shoot. Inflorescences are grouped according to method of branching as (*a*) *indefinite* or *racemose*; (*b*) *definite* or *cymose.*

In (*a*) branching is monopodial. The inflorescence consists of a main axis increasing in length by growth at tip and giving rise to lateral branches bearing flowers. The flowers open in succession from below upwards, or, if inflorescence axis is short and flattened, from outside inwards. Following types of indefinite inflorescence are recognized:– *Raceme*, with a main axis bearing stalked flowers; e.g. lupin, foxglove; *Panicle*, compound raceme, e.g. oat; *Corymb*, a

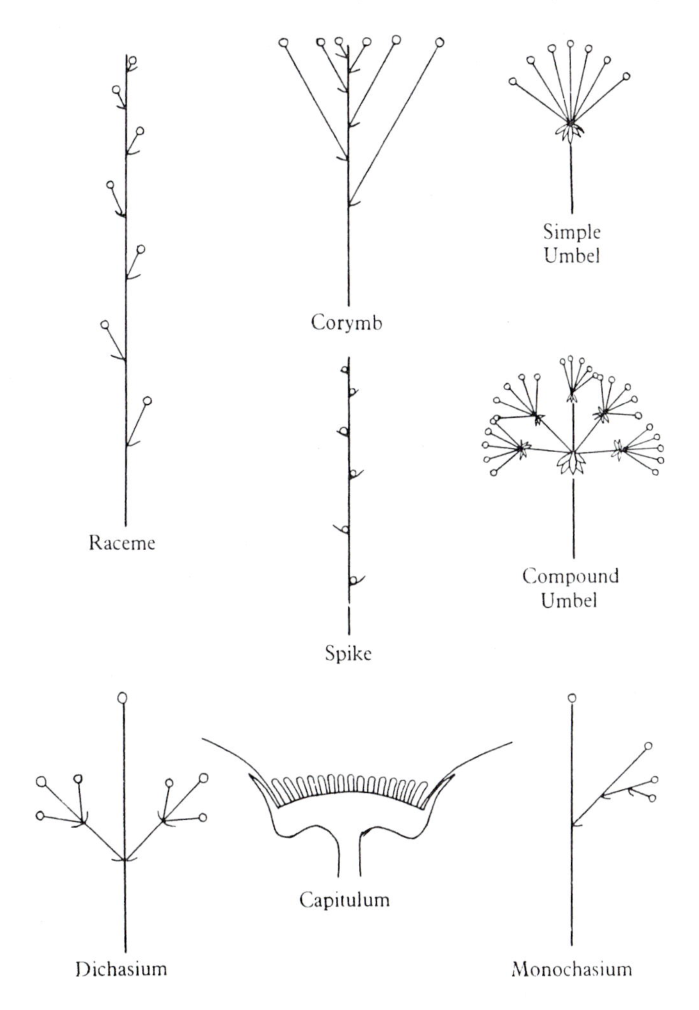

Fig. 5. Diagrams of different types of inflorescence.

raceme with flowers borne at same level, due to elongation of stalks (pedicels) of lower flowers, e.g. candytuft; *Spike*, a raceme with sessile flowers, e.g. plantain; *Spadix*, a form of spike with thick fleshy axis, e.g. cuckoo pint; *Catkin*, a spike of unisexual, reduced flowers, often pendulous, e.g. hazel, birch; *Umbel*, a raceme in which the axis has not lengthened so that the flower stalks arise at the same point, the flowers in a head with the oldest at the outside, youngest in the centre, e.g. carrot, cow parsley; *Capitulum*, in which axis of inflorescence is flattened, laterally expanded, with growing point in centre, and bearing closely crowded, sessile flowers, the oldest at the margin, the youngest in the centre, e.g. dandelion.

In (*b*) branching is sympodial. The main axis ends in a flower and further development takes place by growth of lateral branches, each of which behaves in the same way. The cyme is described as a *Monochasium* when each branch of the inflorescence bears one other branch, e.g. iris, and as a *Dichasium* when each branch gives rise to two other branches, e.g. stitchwort.

Many inflorescences are mixed, partly indefinite, partly definite, e.g. horse-chestnut, a raceme of cymes. See Fig. 5, p. 121.

INFUNDIBULUM. Outpushing from floor of brain (of diencephalon) of embryo vertebrate which meets hypophysis (q.v.) and with it develops into pituitary body (q.v.).

INFUSORIA. Term formerly applied to microscopic organisms found in infusions of organic substances, including various Protozoa and Rotifera; has also been used in a restricted sense for the Ciliophora.

INHIBITION. (Nervous). Prevention, as a result of the influence of nerve impulses, of activation of an effector. May occur in central nervous system: e.g. when a muscle is stimulated by its motor nerve fibres to contract, the tonus of antagonistic muscles is simultaneously relaxed by inhibition of their motor nerve cells (*reciprocal inhibition*); or a reflex can be inhibited by stimulating both the sensory nerve fibres of the reflex and certain other sensory nerve fibres at the same time. Such *central* inhibition is one of the most important ways in which the central nervous system (q.v.) produces its flexible control of activities, by preventing action of effectors in unsuitable circumstances. In Crustacea *peripheral* inhibition occurs, special inhibitory nerve fibres as well as excitatory ones running to the muscles. Similar peripheral inhibition occurs in autonomic system of vertebrates, e.g. vagus nerve inhibits heart beat, sympathetic stimulates it.

INNERVATION. Nerve supply (to a particular organ).

INNOMINATE. (1) Short *artery* arising from aorta and giving rise to a subclavian artery (to fore-limb) and a carotid artery (to head), in many birds and mammals. (2) *I. bone*. Each lateral half of the hip-girdle, when pubis, ilium, and ischium are fused into a single bone, as in adult reptiles, birds, and mammals.

INSECTA (HEXAPODA). Class of Arthropoda containing sprintgails, silverfish, cockroaches, earwigs, termites, flies, green-flies, bugs, lice, fleas, butterflies, beetles, bees, ants. (Some animals commonly referred to as 'insects' do not belong to this group.) Most insects are terrestrial and breathe air by means of tracheae. They have one pair of antennae, three pairs of legs, three distinct parts of the body (head, thorax, and abdomen) and most have wings. See *Apterygota*, *Orthoptera*, *Dermaptera*, *Isoptera*, *Plecoptera*, *Psocoptera*, *Odonata*, *Hemi-*

ptera, Ephemeroptera, Mallophaga, Anoplura, Thysanoptera, Neuroptera, Trichoptera, Lepidoptera, Coleoptera, Hymenoptera, Diptera, Aphaniptera.

INSECTIVORE. Any member of the Insectivora, an order of placental mammals; a primitive insect-eating group, resembling in many respects the Cretaceous ancestors of all placentals. E.g. mole, hedgehog, shrew.

INSECTIVOROUS. Insect eating.

INSTAR. Stage in larval development of an insect, between two ecdyses (q.v.).

INSTINCT. An elaborate pattern of successive reflexes, which occurs as a whole in response to certain stimuli.

INSULIN. Hormone of vertebrates, controlling amount of glucose in the blood; secreted by islets of Langerhans in pancreas. A protein. Secretion is stimulated by high glucose concentration in blood, which insulin decreases by suppressing breakdown of liver glycogen to glucose, and encouraging building up of muscle glycogen from glucose. Antagonized by adrenalin and an anterior pituitary hormone. See *Blood-sugar.*

INTEGUMENT. (Of seed plants), layer enclosing nucellus of ovule (q.v.) and ultimately forming seed coat. Most flowering plants have two integuments, inner and outer.

INTERCALARY. (Of a meristem) situated between regions of permanent tissue, e.g. at base of nodes and leaves in many monocotyledons. Cf. *Apical.*

INTERCELLULAR. Between the cells. *I. bridges* (plasmodesmata in plants), cytoplasmic connexions between cells present in some plant and animal tissues. *I. fluid* (interstitial fluid, tissue fluid), fluid between tissue cells of animals, bathing the plasma-membranes (to be distinguished from blood; see *Internal environment*). *I. material,* is chiefly skeletal, e.g. in animals, reticulin, collagen, elastin, bone salts, matrix of cartilage and connective tissue; in plants cellulose may be regarded as intercellular material, though usually considered part of the cell (q.v.). *I. spaces,* in plants air-filled cavities between walls of neighbouring cells, e.g. in cortex and pith, forming internal aerating system; spaces may be large, rendering tissue light and spongy as in aerenchyma (q.v.). In animals filled with intercellular fluid.

INTERFASCICULAR CAMBIUM. Cambium (q.v.) arising between vascular bundles. Cf. *Fascicular cambium.*

INTERNAL ENVIRONMENT. Medium in which body cells are bathed i.e. the intercellular fluid. In equilibrium with blood-stream in vertebrates, it is normally kept highly constant in composition (homeostasis) e.g. in respect of osmotic pressure, content of individual ions, acidity and alkalinity (pH), and glucose concentration, by sensitive mechanisms which control these conditions in the blood, and hence in the intercellular fluid. Deviations from constancy have rapidly deleterious influence on cells, particularly on brain cells. Constancy does not seem to be so complete in insects, and little is known of other invertebrates.

INTERNODE. (1) Part of plant stem between two successive nodes (q.v.). (2) Part of a nerve fibre between two successive nodes of Ranvier (q.v.).

INTEROCEPTOR. (1) Receptor (q.v.) which detects stimuli arising inside the body. Cf. *Exteroceptor.* (2) In more restricted sense, receptor in lining of gut and respiratory organs, detecting stimuli from substances introduced from outside.

INTERPHASE. Resting stage (q.v.) of nucleus.

INTERSEX. (Zool.). Abnormal individual which is intermediate between the two sexes in characteristics (it may or may not be hermaphrodite, q.v.), having all its cells of identical genetical composition (cf. *Gynandromorph*). May occur through failure of sex-determining mechanism of genes (see *Sex chromosomes*); or through hormonal or other influence during development (see *Free-martin*).

INTERSTITIAL CELLS. Of vertebrate gonads; those which lie between the ovarian follicles or between the testis tubules. Those of the testis probably secrete the testicular sex-hormones (chiefly testosterone, q.v.).

INTERSTITIAL FLUID. See *Intercellular fluid.*

INTESTINE. Alimentary canal between stomach and anus or cloaca. In vertebrates practically all absorption of products of digestion occurs in intestine, and a great deal of digestion; and undigested residue is converted there to faeces. Frequently coiled, so that much longer than body (about ten feet long in man). Internal surface usually increased by folds and projections (see *Spiral valve*, *Villi*). Lined by simple epithelium, containing mucus-secreting glands; its anterior part has digestive enzyme-secreting glands and special absorptive powers. Smooth muscle in walls churns up contents and gradually passes them towards anus. Anterior part receives ducts of large digestive glands, liver and pancreas. Usually anterior part is of smaller diameter than posterior, in which faeces are stored; and in amniotes there is a clear division into *small intestine* and *large intestine*, with a valve and often a *caecum* at their junction. Small intestine is concerned with digestion, and absorption of the products into blood and lymph capillaries (see *lacteals*) in the wall. Large intestine is mainly concerned with preparation of faeces by removal of water, valuable on dry land. Small and large intestines of amniotes are not always homologous with those of anamniotes.

INTRACELLULAR. Within a cell. Cf. *Extracellular.*

INTRASPECIFIC. Amongst members of the same species. *I. selection* natural selection (q.v.) of one kind of individual at the expense of another of the same species, the two kinds directly competing with each other for limited supplies of some requirement. E.g. selection of a character of a male animal (such as large antlers) which gives advantage over other males in reproduction.

INTROGRESSIVE HYBRIDIZATION. Infiltration of genes of one species into genotype of another. When two species come into contact under conditions favouring one or the other, if hybrids are produced they tend to back cross with the favoured (more abundant) species. This process, continually repeated, results in a population of individuals most of which resemble the predominant parent but which possess also some characters of the other parent.

INTRORSE. (Of dehiscence of an anther), towards centre of flower. Cf. *Extrorse*.

INTUSSUSCEPTION. (Bot.). Increase in surface area of cell wall by interpolation of new particles between existing particles of extending wall. Cf. *Apposition*.

INULIN. Soluble polysaccharide, built of fructose molecules, occurring as stored food material in many plants, e.g. members of Compositae, dahlia tubers. Absent in animals.

INVAGINATION. Pushing inwards of part of a sheet of cells so as to form a pocket opening on to original surface. A common process in embryonic development of animals (see *Blastopore*).

INVERSION. Reversal of part of a chromosome so that genes within that part lie in inverse order.

INVERTASE. See *Sucrose*.

INVERTEBRATA. Collective term for all animals which are not members of the vertebrata (e.g. *Amoeba*, sponges, jelly-fishes, worms, snails, flies, starfishes, sea-squirts).

IN VITRO. By derivation, means 'in glass' (vessel). In general, applied to biological processes when they are experimentally made to occur in isolation from the whole organism (which usually means within a glass vessel). E.g. the activities of cells in tissue culture (q.v.) occur *in vitro*. Cf. *In Vivo*.

IN VIVO. Within the living organism. Cf. *In Vitro*.

INVOLUCRE. Protective investment; (1) in thalloid liverworts, scale-like upgrowth of thallus over-arching archegonia; (2) in leafy liverworts and in mosses, group of leaves surrounding sexual organs; (3) in many flowering plants, e.g. family Compositae, group of bracts enveloping young inflorescence.

INVOLUTION. (1) Production of abnormal bacteria, yeasts, etc., e.g. in old cultures. (2) Decrease in size of an organ, opposite of hyperplasia and hypertrophy, e.g. of thymus and other lymphoid tissue after puberty.

IRIS. Structure controlling amount of light admitted to vertebrate eye. Iris is the coloured part of human eye. It is a thin sheet of tissue, forming a pigmented diaphragm in front of lens with central opening (pupil) through which light passes to retina. Attached at outer margin to ciliary body (q.v.). Contains radiating muscles which enlarge pupil and a ring of muscles at its free inner margin which narrows pupil. Light, stimulating retina and optic nerve, reflexly controls size of pupil, narrowing for strong, enlarging for dim light. Accommodation for near vision is also accompanied by narrowing. See Fig. 3, p. 86. A structurally and functionally similar iris occurs in Cephalopod eye.

IRRITABILITY. Responsiveness to change in environment by complex, usually adaptive activity; a universal property of living things. Manifested in, e.g. changes of cell turgor, protoplasmic movements, growth curvatures of plants, nervous activity of animals.

ISCHIUM. Ventral, backward-projecting, part of hip-girdle (q.v.). Ischia bear the weight of a sitting primate. See Fig. 10, p. 216.

ISLETS OF LANGERHANS. Groups of cells, scattered throughout pancreas of gnathostome vertebrates, which secrete insulin (q.v.).

ISOBILATERAL. (Of leaves), having same structure on both sides. Characteristic of leaves of monoctyledons, e.g. iris, where leaf-blade is more or less vertical and the two sides are equally exposed. Below epidermis on both sides of the leaf, lying regularly side by side with their long axes at right-angles to it, are elongated cells of palisade mesophyll and between them, loosely arranged cells of spongy mesophyll. Cf. *Dorsiventral.*

ISOGAMY. Condition in which gametes are similar, i.e. not differentiated into male and female; found in some green Algae, Fungi, and Protozoa, but uncommon. Cf. *Anisogamy.*

ISOMORPHIC. Used of alternation of generations (q.v.), particularly in Algae, meaning generations vegetatively identical. Cf. *Heteromorphic.*

ISOPODA. Order of Crustacea, including woodlice, water-slaters, pill-bugs, etc.

ISOPTERA. Termites (white ants); order of exopterygote insects. Live in huge colonies with elaborate system of castes, resembling that of true ants, though independently evolved.

ISOTONIC. (Of a solution), having a concentration such that it neither gains nor loses water by osmosis (q.v.) when separated by a membrane from a specified other solution. When the membrane and the other solution are unspecified, they are taken to be the plasma-membranes and interior of cells respectively. Cf. *Hypertonic*, *Hypotonic.*

J

JEJUNUM. Part of small intestine succeeding duodenum and preceding ileum in mammals. Main absorptive region. Larger diameter and larger villi than rest of small intestine.

JUGULAR. Main vein returning blood from head (particularly brain) in vertebrates. See *Cardinal veins.*

JURASSIC. Geological period (q.v.); lasted approximately from 180 till 135 million years ago.

K

KARYOKINESIS. Mitosis (q.v.).

KARYOTYPE. Appearance (size, shape, and number) of the set of chromosomes of a somatic cell.

KATABOLISM. Catabolism (q.v.).

KATHEPSIN. Cathepsin (q.v.).

KEEL (CARINA). Thin plate-like projection of bone, from ventral surface of breast-bone (sternum) of birds and bats, to either side of which the powerful wing-muscles are attached.

KERATIN. Tough fibrous protein containing much sulphur, occurring in epidermis of vertebrates, forming resistant outermost layer of skin, and also hair, feathers, horny scales, nails, claws, hooves, and outer coating of horns of cows, sheep, etc. *Keratinization* is transformation of epidermal cells into keratin, which occurs in formation of above structures.

KIDNEY. Organ of excretion and/or water regulation in vertebrates, consisting of numerous nephrons (q.v.) and their blood supply. See also *Coelomoduct*, *Nephridium*.

KILOCALORIE. See *Calorie*

KINAESTHETIC. Detecting movement. Refers, e.g. to sense-organs of vertebrate muscles, tendons, and joints. See *Proprioceptor*.

KINESIS. Locomotory movement of an organism or cell in response to a stimulus, such that the speed, or frequency of turning, depends on strength of stimulus; but stimulus does not control direction of movement. Cf. *Taxis*.

KREBS CYCLE (CITRIC ACID CYCLE). Complex cycle of enzyme-controlled reactions by which pyruvic acid is broken down in presence of oxygen to carbon dioxide. The cycle provides the final step in oxidation of carbohydrates, since glycogen and glucose are broken down to pyruvic acid. It also deals with the final phases of fat oxidation and is concerned in synthesis of some amino acids. The cycle seems to be particularly associated with the mitochondria.

L

LABIAL. (Adj.). Of the lip.

LABIUM. Lower lip of insect. Consists of the paired appendages of one segment fused together in mid-line.

LABRUM. Plate of exoskeleton, hinged to the head above the mouth, which forms the upper lip of insects.

LABYRINTH, MEMBRANOUS. See *Ear, inner*.

LABYRINTHODONTIA. The main group of fossil Amphibia. Late Devonian to Triassic. Contains forms closely related to certain fish (Crossopterygii) and others transitional to reptiles.

LAC. Resinous exudation covering females of an insect, *Tachardia* (Hemiptera). The main constituent of shellac.

LACERTILIA (SAURIA). Lizards. A sub-order of Squamata (sometimes a separate order), class Reptilia. Unlike the other squamate sub-order, the Ophidia (snakes), lizards have normal-sized jaw gape, movable eyelids, and eardrums; and most have normal limbs.

LACHRYMAL GLAND. Tear gland of eye of tetrapod vertebrate. Lies beneath upper eyelid of man and other mammals. Continually secretes small amounts of sterile, and slightly antiseptic, tears which keep cornea moist. Tears drain into nose through *lachyrmal duct* from inner corner of eye.

LACTASE. See *Lactose.*

LACTATION. Production of milk. See *Mammary gland*, *Lactogenic hormone*.

LACTEALS. Lymph vessels draining villi (q.v.) of intestine in vertebrates. Into them passes fat absorbed from intestine, and the fat globules give milky appearance to contained lymph.

LACTIC ACID. Organic acid formed as product of splitting up of glucose (glycolysis), which is an essential process in utilization of energy of food by many animal cells, and particularly during contraction of vertebrate striped muscle. Also formed in metabolism of many bacteria, e.g. from lactose in souring of milk (hence its name).

LACTOGENIC HORMONE (PROLACTIN). Hormone of anterior pituitary; in mammals probably largely responsible for onset of milk production in mother after parturition; also stimulates crop-milk (q.v.) production in pigeons; and is probably responsible for maternal, broody and related behaviour in mammals, birds, amphibia and fish.

LACTOSE. Sugar (disaccharide, with twelve carbon atoms) occurring in mammalian milk. Compound of a molecule of glucose and a molecule of galactose (into which it is split by intestinal enzyme *lactase*).

LAMARCKISM. View that acquired characters (q.v.) are inherited, first systematically propounded by Lamarck (1744-1829). Lamarck however was principally concerned with a supposed evolutionary effect of *habits* acquired during the life of an animal. The rather different modern theories of the inheritance of responses to environmental influence are hence often called Neo-Larmarckism.

LAMELLA. Thin layer or plate-like structure. (1) One of spore-bearing gills in fruit-body of mushroom or related fungus, attached to underside of cap and radiating from centre to margin. (2) Of bone. One of thin layers (roughly five microns thick) in which the calcified matrix (q.v.) of bone occurs. Frequently disposed concentrically around Haversian canal (q.v.), or parallel to surface of bone.

LAMELLIBRANCHIATA (PELECYPODA). Bivalves; class of molluscs with bivalve shells, including mussels, clams, oysters. Aquatic; elaborate mechanism of feeding by ciliary currents.

LAMINA. Leaf-blade.

LAMP SHELLS. Brachiopoda (q.v.).

LANUGO. Crop of very fine hair covering human foetus, which disappears before birth.

LARVA. The pre-adult form in which some animals hatch from the egg; capable of fending for itself though usually in a way different from adult; but usually incapable of sexual reproduction (see *Paedogenesis*); and distinctly different from sexually mature adult in form. Turns into adult, usually by a rather rapid metamorphosis (q.v.). E.g. tadpole of frog, caterpillar of butterfly; and many marine invertebrates (Echinodermata, Mollusca, Annelida) have planktonic, transparent, ciliated larvae quite unlike adults (see *Trochophore*, *Veliger*).

LARYNX. Dilated region of upper part of trachea of tetrapods, just at its junction with pharynx. 'Adam's apple' of man. Has plates of cartilage in its walls, movable by muscles which open and close glottis (q.v.). In some tetrapods, including most mammals, a dorso-ventral fold of membrane (vocal cord), situated within larynx, projects into its lumen from each side wall. Vibration of cords produces vocal sounds. Movement of the cartilage plates varies stretch of vocal cords, and hence alters pitch of voice. See *Syrinx*.

LATENT PERIOD. Time between application of stimulus and first detectable response in an irritable tissue (e.g. muscle or nerve, where it is a fraction of a millisecond). Also applied to a nervous reflex; but term *reaction time* is more often used for reactions of a whole animal or plant.

LATERAL LINE SYSTEM (ACUSTICO-LATERALIS SYSTEM). System of sense organs, present in aquatic vertebrates (Agnatha, fish, and aquatic Amphibia), in pores or canals arranged in a line down each side of the body, and in complicated pattern of lines on head. Detects pressure changes including vibrations in water. Canals in dermal bones of head lodge the system, and are important in establishing homologies of these bones in different fish. Inner ear is probably a modified part of system. Supplied entirely by cranial nerves.

LATERAL PLATE. Mesoderm on lateral and ventral sides of vertebrate embryos. Unsegmented (cf. *Somitic Mesoderm*). Within it the perivisceral coelom arises. Forms peritoneum, smooth muscle of gut, heart, pericardium, muscle and bone of tetrapod limb, etc.

LATEX. Fluid produced by a number of flowering plants characteristically exuding from cut surfaces as a milky juice, e.g. dandelions, lettuce. Contains a number of substances including proteins, sugars, mineral salts, alkaloids. oils, caoutchouc, etc., and rapidly coagulates on exposure to air. Function of latex not clearly understood but thought to be concerned in nutrition and in protection and healing of wounds. Latex of a number of species is collected and used in manufacture of several commercial products, most important being rubber.

LATIMERIA. The only known living representative of the great group of fossil fish, Crossopterygii. A coelacanth.

LEAF BLADE. Thin, flattened, conspicuous portion of leaf, seat of photosynthesis (q.v.); may be simple, consisting of one piece, or compound, divided into separate parts or leaflets.

LEAF GAP. Localized region in vascular cylinder of stem immediately above point of departure of a leaf trace (leaf trace bundle) (q.v.) where parenchyma is differentiated instead of vascular tissue. In some plants where there are several leaf traces (leaf trace bundles) to a leaf these are associated with a single leaf gap. Leaf gaps are characteristic of ferns, gymnosperms and angiosperms.

LEAF-SCAR. Scar marking position where a leaf was formerly attached to the stem.

LEAF-SHEATH. Base of leaf modified to form a sheath round the stem, e.g. grasses, sedges.

LEAF-TRACE. (1) Vascular bundle extending between vascular system of stem and leaf base; where more than one occurs each constituting a leaf trace; (2) Vascular supply extending between vascular system of stem and leaf base, consisting of one or more vascular bundles, each known as a *leaf trace bundle.*

LECITHIN. Fatty substance (lipine) containing glycerol, fatty acid, choline, and phosphoric acid, present in all animal and plant cells.

LECTOTYPE. Specimen selected from the original material to serve as the type specimen (q.v.) when this was not designated at the time of publication or is missing. Cf. *Neotype.*

LEGUME. (1) A pod. Fruit of members of family Leguminosae (peas, beans, gorse, laburnum, etc.). Dry fruit formed from a single carpel that liberates its seeds by splitting open along both sutures into two parts or valves. (2) Used by agriculturists for a particular group of fodder plants belonging to this family, clovers and allied plants.

LEMMA (INFERIOR PALEA). Glume-like bract of grass spikelet bearing in its axil a flower, which, with palea (q.v.), it encloses.

LEMUR. One of a group of primitive primates, found in Old World tropics.

LENS. Of eye of vertebrates, transparent structure just behind pupil, lying in aqueous humour, attached by collagen fibres to ciliary body. In land-living forms, including man, not as important as cornea in the refraction which produces image on retina; function is accommodation (q.v.) by change of shape. In fish, spherical, and responsible for both refraction and accommodation. In embryonic development of vertebrate originates from epidermis, as an induction (q.v.) by retina. See Fig. 3, p. 86. Lenses and lens-like structures occur in eyes of many invertebrates.

LENTICEL. Small raised pore, usually elliptical in shape, developing in woody stems when epidermis is replaced by cork; packed with loosely arranged cells and allowing exchange of gases between interior of stem and atmosphere.

LEPIDOPTERA. Butterflies, moths. Order of endopterygote insects. Two pairs of large wings; wings and body covered with scales; larva a caterpillar. Adults feed on nectar of flowers using a highly specialised proboscis; larvae usually feed on plants, and include some serious pests.

LEPIDOSIREN. South American lungfish (member of Dipnoi, q.v.).

LEPTOCEPHALUS. Oceanic larva of European eel. Migrates more than 2000 miles across Atlantic from breeding-ground of eels near West Indies to reach European fresh-waters, where becomes adult. Quite transparent.

LETHAL GENE. Gene which kills individual bearing it; or, if inherited, kills individuals homozygous, but not heterozygous, for it.

LEUCOCYTE. White blood cell (q.v.).

LEUCOCYTOSIS. Presence of unusual number of white blood cells in blood, a response to infection or tissue damage.

LEUCOPLAST. Colourless plastid (q.v.) in which starch is deposited; found in plant tissues not normally exposed to light.

LIANES. Climbing plants found in tropical forests, with long, woody, rope-like stems of anomalous structure.

LICHENES. Lichens, subdivision of Thallophyta. Dual organisms formed from symbiotic association of two plants, a fungus and an alga. The fungus partner is usually an Ascomycete, sometimes a Basidiomycete, the algal partner a green (Chlorophyceae) or blue-green (Cyanophyceae) alga. A cosmopolitan group of plants occurring on tree trunks, old walls, on the ground, exposed rock, etc., and providing the dominant flora in large areas of mountain and arctic regions where few other plants can live. Lichens play an important part with liverworts and mosses in the primary colonization of bare areas.

Lichens are *crustose*, forming a thin, flat crust on the substratum, *foliose*, flat with leaf-like lobes, or *fructicose*, upright, branched forms. Very slow growing and vary greatly in size, e.g. from a millimetre to several metres across. Reproduce vegetatively by *soredia*, groups of fungal hyphae enclosing algal cells, cut off and dispersed in large numbers. Sexual reproduction is confined to fungus partner and in majority of lichens it is by the development of apothecia or perithecia. On germination of the ascospores new lichen plants are formed if the algal partner happens to be present but in its absence the fungus dies.

In arctic regions certain lichens are valuable as a source of food, e.g. Iceland moss, reindeer moss. Other lichens provide dyes, e.g. *Roccella*, providing litmus.

LIFE CYCLE. Progressive series of changes undergone by an organism or lineal succession of organisms, from fertilization to death, or to the death of that stage producing the gametes which begin an identical series of changes. In vertebrates this would simply be from union of gametes to death of the resulting individual; but in many plants and animals there is a succession of individuals, with sexual or asexual reproduction connecting them, in the entire cycle, e.g. in flukes (q.v.). See *Alternation of generations.*

LIGAMENT. Strong band of collagen connecting the two bones at a joint. Helps restrict movement to that provided for by shape of joint, preventing dislocation.

LIGNIN. Complex aromatic compound, chemistry of which is not fully understood, which is deposited in cell walls of sclerenchyma, xylem vessels and tracheids, making them strong and rigid. Forms 25–30 per cent of the wood of trees.

LIGULE. Membranous outgrowth arising (1) from junction of leaf-blade and leaf-sheath in many grasses and (2) from base of the leaves of certain lycopods.

LIMNOLOGY. The study of fresh waters and their inhabitants.

LIMULUS. King-crab, an aquatic arachnid, only living representative of its order (Xiphosura). Apparently identical animals existed in the Triassic.

LINGUAL. (Adj.). Of the tongue.

LINKAGE. Association of two or more non-allelomorphic genes, so that they tend to be passed from generation to generation as an inseparable unit, and fail to show independent assortment (q.v.). Due to their being in same chromosome. Separation of such genes into different chromosomes occurs from time to time as a result of crossing-over (q.v.). The nearer such genes are to each other on a chromosome the less often they are separated by crossing-over and the more *closely linked* they are said to be. All the genes in one chromosome form one *linkage-group.*

LINOLEIC ACID (VITAMIN F). Unsaturated fatty acid required by various insects and mammals, probably including man.

LIPASE. Enzyme which splits esters of fatty acids, e.g. true fats, into alcohol and acid.

LIPID (LIPIDE). See *Fat.*

LIPIN. Synonymous with lipine (q.v.); or sometimes with fat (q.v. meaning (1) or (2)).

LIPINE. Fat (q.v. meaning (1)) containing nitrogen, and often phosphorus (phospholipin) or sulphur. Always complicated compound, e.g. lecithin (q.v.).

LIPOID. (1) Substance resembling fat (q.v.) in solubility, but not containing fatty acid (e.g. steroid, carotene, terpene); (2) Substance, resembling fat in solubility, which may contain fatty acid (e.g. phospholipin); but excluding neutral fats. (3) Occasionally, fat (q.v. meaning (1) or (2)).

LITHOSERE. Type of sere (q.v.).

LITTORAL. Inhabiting bottom of sea or lake near shore, roughly within a depth to which light and wave action reach. For sea, usually taken as between high tide mark and 200 metres (i.e. approximately to limit of continental shelf). For lakes, approximately down to 10 metres.

LIVER. Gland whose duct opens into gut, usually with digestive function. Not homologous in different phyla. Liver of vertebrates has many functions, e.g. secretion of bile (q.v.); storage of glycogen (see *Blood-sugar*).

LIVER-FLUKE. Of sheep, cattle, etc., fluke (q.v.) which lives in bile ducts and causes liver-rot, disease which leads to great losses. Common in wet meadows where the intermediate host, a water snail, flourishes and infects the pastures with cercaria larvae.

LIVERWORTS. Hepaticae (q.v.).

LOCULICIDAL. (Bot.). Describing dehiscence of multilocular capsule by longitudinal splitting along dorsal suture (midrib) of each carpel, e.g., iris. Cf. *Septicidal.*

LOCUS. A position in a chromosome at which there is always one, and only one, gene of the same kind, or one of a set of genes allelomorphic to each other. Homologous chromosomes contain identical sets of loci in the same linear order. Loci pair in meiosis.

LODICULES. Reduced perianth of grass flowers; two small, scale-like structures below ovary which at time of flowering swell up, forcing open the enclosing bracts (pales), exposing stamens and pistil.

LOMENTUM. Form of legume fruit (q.v.) constricted between the seeds and breaking into one-seeded portions when ripe, e.g. birds-foot.

LONG-DAY PLANT. See *Photoperiodism.*

LUMBAR VERTEBRAE. Vertebrae of the waist region; without ribs; between the rib-bearing thoracic and the sacral vertebrae. Present in most amniotes.

LUMEN. Cavity. Space within a tube, sac, or cell-wall.

LUNG. Organ for breathing air. (1) In vertebrates. Embryologically a diverticulum of the gut. Present in early fishes before origin of Amphibia, probably as an accessory breathing organ adapted to oxygen-poor fresh waters in which earliest vertebrates probably lived. A functional lung occurs in some living fishes (*Polypterus*, Dipnoi). In others it has evolved into swim-bladder (q.v.). Present in almost all amniotes. See *Bronchus*, *Bronchiole*, *Alveólus.* (2) In land molluscs, highly vascular part of mantle.

LUNG BOOK. Respiratory organ of some air-breathing arachnids (spiders and scorpions); consists of projections, containing blood and arranged like leaves of a book, in a depression of the body wall.

LUTEAL PHASE. See *Oestrus Cycle.*

LUTEAL TISSUE. Tissue which fills cavity of a ruptured Graafian follicle, constituting a corpus luteum (q.v.). Derived from follicle cells. Secretes progesterone.

LUTEINIZING HORMONE (L. H.). Gonadotrophic hormone (q.v.) secreted by anterior lobe of pituitary in mammals, which stimulates ovulation and formation of corpora lutea (q.v.) in females and secretion of androgen in males. See *Follicle-stimulating hormone.*

LYCOPODIALES. Club mosses. Order of Pteridophyta with fossil history extending back to Palaeozoic. Attained their highest development in Carboniferous and included tree-like forms, e.g. *Lepidodendron, Sigillaria.* Existing representatives small, evergreen plants with upright or trailing stems bearing numerous small leaves. Sporangia borne singly in axils of leaves, sporophylls occurring in groups at intervals along stem or forming terminal cones. Homosporous, e.g. *Lycopodium,* with small prothalli, wholly or partly subterranean, mycotrophic; or heterosporous, with reduced prothalli remaining largely enclosed by spore wall, e.g. *Selaginella.*

LYMPH. Fluid drained by lymph vessels from intercellular spaces; its water-content ultimately derived from blood by filtration through capillary walls. Colourless, containing small amount of protein, and varying numbers of cells, chiefly lymphocytes. See *Lymphatic system.*

LYMPHATIC (L. VESSEL). Lymph-carrying vessel, lined by smooth endothelium like a blood-vessel; but always with thin walls, which in larger vessels contain smooth muscle and connective tissue. Many have valves. The smallest vessels are *lymphatic capillaries.* In Anura some of the vessels are enlarged into very big *lymph spaces.* See *Lymphatic system, Lacteal.*

LYMPHATIC SYSTEM. System of fluid-containing tubes present in vertebrates. A blindly-ending meshwork of small tubes (lymph capillaries) permeates most tissues (not nervous system), and connects up into ever larger vessels (but not usually larger than 2-3 mm. diameter) which finally join venous system usually near the heart. Lymph (q.v.) is drained from the tissues into the blood by this system. Because of high permeability of lymph capillaries, particles from the intercellular spaces such as colloids, tissue debris, or invading bacteria, which cannot get through blood capillary walls, are carried away with the lymph. In the course of the lymphatic vessels in some vertebrates are lymph nodes, which filter out and destroy bacteria in the lymph, and supply lymphocytes to it. The lymph moves by contractility of vessels; or by squeezing of lymphatic vessels by neighbouring skeletal muscles, the larger vessels having valves which ensure flow in one direction; or (except in birds and mammals) by lymph hearts.

LYMPH HEART. Enlarged part of lymphatic vessel with muscular pulsating wall, present in many vertebrates, but not in birds or mammals. Pumps lymph.

LYMPH NODE (LYMPH GLAND). Organ on the course of a main lymphatic, consisting of lymphoid tissue (q.v.). Its macrophages remove foreign bodies from lymph. If these foreign bodies are bacteria, it may become inflamed ('swollen glands'). It is an important source of antibodies. Occurs in mammals, and, not so well-developed, in birds; not in other vertebrates.

LYMPHOCYTE. Kind of white blood cell (q.v.) of vertebrates. Rather small (6-8 microns diameter in man) but with relatively large spherical nucleus. Forms 20-25 per cent of white blood cells in man. Non-phagocytic, actively amoeboid, suspected of producing or carrying antibodies. Lymphocytes are continuously made in lymphoid tissue.

LYMPHOID TISSUE (LYMPHATIC TISSUE). Tissue of vertebrates which produces lymphocytes (q.v.) by division of some of its cells; it also contains macrophages (q.v.). Widely distributed in body. Occurs especially in spleen, lymph nodes, thymus, tonsils.

LYSIGENOUS. (Of secretory cavities in plants), originating by dissolution of secreting cells, e.g. oil-containing cavities in orange peel. Cf. *Schizogenous*.

M

MACROGAMETE (MEGAGAMETE). Female gamete, distinguished from male gamete by larger size and/or by structure. See *Anisogamy*.

MACRONUCLEUS (MEGANUCLEUS). Members of Ciliophora have two kinds of nuclear material, one constituting the large macronucleus, which divides amitotically and is apparently concerned with the vegetative processes of the organism; and the other the smaller *micronucleus*, which divides mitotically and provides the gametes during conjugation. The macronucleus disappears during conjugation, and is reconstituted from the zygote nucleus, i.e. from micronuclear material. More than one macronucleus and micronucleus may occur in each organism.

MACROPHAGES. Phagocytic cells widely distributed in vertebrate body. Occur in all connective tissue, where after injury they become active in removing the debris. The resting, connective tissue forms are often called *histiocytes*. Occur also (forming the reticulo-endothelial system) in contact with lymph (in lymph-nodes) or blood (in bone-marrow, spleen, liver), removing from these fluids any foreign particles. Can be readily demonstrated by injecting an animal with Indian ink or certain dyes, the particles of which are engulfed by, and mark, the macrophages. Move by membrane-like pseudopodia. See *Reticulo-endothelial system*, *Monocyte*.

MACROPHAGOUS. (Of animals). Feeding on pieces of food large relative to their own size. Feed at intervals. All land animals are macrophagous. Cf. *Microphagous*.

MACROSPORANGIUM. Megasporangium (q.v.).

MACROSPORE. Megaspore (q.v.).

MACROSPOROPHYLL. Megasporophyll (q.v.).

MACULA (OF EYE). See *Fovea*.

MAGGOT. Worm-like larva of certain insects (dipteran flies) with no appendages and without well-marked head.

MALARIA PARASITE. Genus of sporozoan Protozoa (*Plasmodium*), causing malaria in man.

MALLEUS. Mammalian ear-ossicle (q.v.) representing articular of other vertebrates.

MALLOPHAGA. Biting lice, bird lice. Order of small exopterygote insects. Live on birds as scavengers, eating bits of feather, skin, etc.; they cannot pierce and suck. Cf. *Anoplura*.

MALPIGHIAN BODY (M. CORPUSCLE). Filtering unit of vertebrate kidney; a Bowman's capsule with its glomerulus.

MALPIGHIAN LAYER. Layer of epidermis next to dermis; in it active cell-division goes on; some of the new cells thus formed leave the layer, becoming keratinized and replacing the outer cells of the epidermis which are continually wearing away. Contains most of the melanin responsible for dark skin.

MALPIGHIAN TUBULES. Tubular glands, excretory in function, opening into anterior part of hind gut of insects, arachnids, and myriapods.

MALTASE. See *Maltose*.

MALTOSE. Sugar (disaccharide, with twelve carbon atoms) formed as a result of starch breakdown. Occurs, e.g. in germinating seeds (such as malt) and during digestion. A maltose molecule is a compound of two molecules of glucose (into which it is split by enzyme *maltase*).

MAMMA. Mammary gland (q.v.).

MAMMAL. Any member of the Mammalia, a class of tetrapod vertebrates. E.g. man, dog, whale. Contains three living sub-classes, Monotremata, Marsupialia, and Placentalia, the latter two subclasses containing also numerous fossils; and some other, imperfectly known, fossil groups dating as far back as Triassic. Peculiar to mammals are: hair, milk secretion; diaphragm used in respiration; lower jaw of single pair of bones; presence of only left systemic arch (q.v.); three bones (auditory ossicles) in each middle ear connecting ear-drum and inner ear.

MAMMARY GLAND. Milk-producing gland on ventral surface, peculiar to female mammals. Probably represents modified sweat-glands. Growth and activity is affected by gonad hormones (q.v.) and state of the gland usually varies with oestrous cycle (q.v.); actual milk production depends on pituitary lactogenic hormone (q.v.).

MANDIBLE. (1) Of vertebrates, lower jaw. (2) Of insects, Crustacea, and Myriapoda, one of the pair of mouth-parts (q.v.), which usually does most of the work of biting and crushing food.

MANTLE. Of molluscs, fold of skin covering whole or part of body. Outer surface secretes shell. Protects feeding and/or respiratory organs, which lie in space between body and mantle. A similar structure is found in brachiopods.

MANUS. Hand. Carpus, metacarpus, and digits of tetrapod fore-limb. See *Pentadactyl Limb*, *Pes*.

MARSUPIALIA (METATHERIA, DIDELPHIA). A sub-class of Mammalia, living in Australia and North and South America. Like placental mammals, the young develop inside uterus of mother, but usually with a placenta (q.v.) connected with the yolk sac instead of with the allantois. Born in a very undeveloped state, and usually sheltered in a pouch (marsupium) which contains the teats of the milk glands. Includes opossum and a few other species in New World, kangaroo and many other animals of Australia; much more widespread as fossils.

MARSUPIUM. Pouch of many marsupials and *Echidna*. Fold of skin supported by epipubic bones of hip girdle, forming pouch containing mammary glands, into which the new-born (or eggs of *Echidna*) are placed. Young are attached to teats in marsupials; lick milk in *Echidna* which has no teats.

MASTIGOPHORA. Flagellata (q.v.).

MASTOID PROCESS. Excrescence of human skull (of periotic bone) just behind ear, containing air-spaces which communicate with middle ear.

MATRIX. (Zool.). Intercellular substance in which cells are embedded, e.g. in cartilage, or connective tissue. Bone matrix is intercellular substance in which bone salts are deposited.

MATURATION OF GERM-CELLS. (Zool.). Process of development of mature sperm and ova, from normally sized and proportioned precursor cells, which occurs in testis and ovary respectively. It includes reduction division or meiosis of the nucleus. which halves number of chromosomes; and extensive changes in cytoplasm, quite different in the two sorts of gametes. See *Oocyte*, *Spermatocyte*.

MAXILLA. (1) One of the (dermal) bones of the upper jaw of vertebrates (and of the face in man) which carries all of the upper teeth, except incisors. (2) Sometimes used for the whole upper jaw of vertebrates. (3) One of a pair of mouth-parts in insects. Crustacea, and Myriapoda, lying behind the mandibles (q.v.); assists in eating.

MEATUS. A passage. E.g. external auditory meatus (see *Ear, outer*).

MECKEL'S CARTILAGE. Bar of cartilage, one forming each side of lower jaw of embryo gnathostome vertebrate, and of adult elasmobranch. In most other adult vertebrates it becomes reduced, and partly ossified as articular, which (except in mammals, see *Ear ossicles*) always forms hinge of lower jaw, rest of lower jaw being composed of dermal bones. Represents part of a visceral arch (q.v.).

MECONIUM. Contents of intestine of mammalian foetus. Derived from secretions of glands discharging into gut, and swallowed amniotic fluid.

MEDIAN. Situated in or towards the plane which divides a bilaterally symmetrical organism or organ into right and left halves.

MEDIASTINUM. Median space in the chest of mammals; a narrow dorsoventrally extended cleft between the two pleural cavities (q.v.), containing the heart in its pericardium, aorta, trachea, oesophagus, thymus, etc.

MEDULLA. Central part of an organ. (1) *of plants*, pith; central core of usually parenchymatous tissue in those stems in which the vascular tissue is in the form of a cylinder; functioning in food storage. May occur in some roots where central tissue develops into parenchyma instead of xylem. (2) *Adrenal m.* see *Adrenal gland.* (3) Of brain, see *Medulla oblongata.*

MEDULLA OBLONGATA. Often called simply medulla. Most posterior part of the brain of vertebrates, merging into the spinal cord. Thick-walled and floored, but with thin roof; its internal cavity is fourth ventricle. Primitively concerned with co-ordination of impulses from lateral line, ear, taste, and touch receptors. Contains important centres for respiratory movement, control of blood-vessels and heart, etc. Glossopharyngeal and vagus nerves arise from it.

MEDULLARY RAY. Thin, vertical plate of parenchyma cells, one to several cells wide, running radially through tissues of stele (q.v.). Medullary rays are either *primary*, passing from pith (medulla) to cortex between the primary vascular bundles, or *secondary*, formed from cambium during process of secondary thickening and ending blindly in secondary xylem and phloem. Since these latter have no connexion with the pith (medulla) they are called by some authorities *vascular* (phloem or xylem) *rays.* Large numbers of medullary rays are present in the vascular cylinder and they form system of radiating plates of living tissue connected with each other by living parenchyma cells. The function of the medullary rays is in storage and radial conduction of synthesized food material.

MEDULLATED NERVE FIBRE. Synonym of myelinated nerve fibre. See *Nerve fibre*, *Myelin.*

MEDUSA. Free swimming form of coelenterate (of subphylum Cnidaria, e.g. jelly-fish). Bell or umbrella shaped, swimming by pulsations of body. Generally do not reproduce by budding. Reproduce sexually, fertilized eggs giving rise to polyps. Are themselves produced asexually by polyps (q.v.). See *Alternation of generations.*

MEGAGAMETE. Macrogamete (q.v.).

MEGANUCLEUS. See *Macronucleus.*

MEGASPORANGIUM (MACROSPORANGIUM). Sporangium within which megaspores are formed. In flowering plants known as the ovule. Cf. *Microsporangium.*

MEGASPORE (MACROSPORE). Larger of the two kinds of spore produced by heterosporous ferns; the first cell of female gametophyte generation of these plants and of seed plants. In flowering plants becomes embryo sac (q.v.). Cf. *Microspore.*

MEGASPOROPHYLL (MACROSPOROPHYLL). Leaf or modified leaf bearing megasporangia. In flowering plants, the carpel. Cf. *Microsporophyll.*

MEIOSIS (REDUCTION DIVISION). Two successive cell divisions of special kind, starting in a diploid cell. Both divisions resemble mitoses, but the chromosomes are duplicated only once in the whole process, so that the number of chromosomes present in each of the four daughter-cells is half that of diploid cell (i.e. the daughter-cells are haploid). Meiosis occurs at some time in the life cycle of all sexually reproducing organisms, because gametes must be haploid to compensate for the chromosome doubling produced by fertilization. In animals, and some algae, it occurs during formation of gametes. In many Fungi and green Algae and in Sporozoa, it occurs immediately after fertilization or on germination of zygote. In most other plants it occurs some time after fertilization but some time before gamete formation, during formation of spores (see *Alternation of generations*).

The course of meiosis is extremely similar wherever it occurs. Chief differences from two successive mitoses are as follows (Cf. *Mitosis*; Figs. 6A, 6B, pp. 147, 148). When chromosomes first appear (in prophase) they are *single* (in mitosis they have already doubled). The two homologous members of each pair of chromosomes then associate closely side by side, corresponding loci adhering together, a process called *pairing* or *synapsis*; each associated pair is a *bivalent*; Fig. 6B (2b). The bivalents shorten and thicken. Each individual chromosome then duplicates, so that each bivalent becomes four *chromatids* (2c). The two chromatids derived from one chromosome remain paired but they separate from the other two chromatids derived from the homologous chromosome. At certain places, however, they are held together by interchanges (chiasmata) occurring between chromatids derived from homologous chromosomes. Chiasmata are the cause of crossing-over (q.v.) of genes. There are usually one or more chiasmata per chromosome pair per meiosis. In a chiasma two chromatids, one from each of the original chromosomes, have apparently broken at corresponding places; and the broken ends of one chromatid have fused with the broken ends of the other, Fig. 6B (2c). Where there is more than one chiasma in a single bivalent they may involve different pairs of chromatids, but always one from each of the original chromosomes. At anaphase two of the four chromatids from each bivalent go to one pole of the spindle, and the other two to the other pole. The chromatids go in pairs because the spindle attachment of each original chromosome has not yet duplicated, so that two chromatids are united at that point. Owing to the effects of chiasmata, however, the united chromatids are not usually derived throughout their length from the same chromosome (though they always are in the immediate neighbourhood of the spindle attachment) but are mixtures of one or more pieces from each of the original chromosomes. It is a matter of chance, and quite uninfluenced by behaviour of other bivalents, which spindle attachment with its chromatids, goes to which pole of the spindle (Fig. 6B (4)).

After anaphase there may be a short telophase (5) and resting stage, or the second meiotic division may follow immediately, usually in both the daughter-cells formed by the first division. It is like a mitosis, except that it starts prophase with only half the normal number of chromosomes, each already divided into two chromatids at the previous prophase. These two chromatids separate at anaphase (4′), one going to each daughter cell. Again, the distribution of the chromatids is a matter of chance, except that the two from each chromosome must go to opposite poles. On this chance distribution of the chromatids at both anaphases depends the law of Independent Assortment (q.v.). Each daughter-cell therefore eventually receives only the haploid number of chromosomes (5′). The reduction from the diploid number of the original cell to the haploid number of the four products is one of the bases of genetic segregation (q.v.).

MELANIN. Dark-brown pigment present in many animals which in different concentrations gives brown and yellow coloration. Colour of human hair is mainly due to melanin.

MELANOPHORE. Chromatophore (q.v.) containing melanin as pigment.

MEMBRANE BONE. Dermal bone (q.v.).

MEMBRANOUS LABYRINTH. See *Ear, inner*.

MENDELISM. Science of the behaviour of genes (q.v.) in inheritance studied by breeding experiments.

MENDEL'S LAWS. First Law, of Segregation (q.v.). Second Law, of Independent Assortment (q.v.).

MENINGES. Membranes covering the vertebrate central nervous system: dura mater (q.v.) outside pia-arachnoid (q.v.).

MENSTRUAL CYCLE. Modified oestrous cycle (q.v.) of catarrhine primates (Old World monkeys, anthropoid apes, and man) which has special feature of sudden destruction of mucosa of uterus at end of luteal phase of cycle, producing bleeding; and absence of any well-marked period of oestrus (q.v.).

MENTAL PROMINENCE. Chin. Projection at front of lower jaw bone (dentaries) of man. Not to be confused with intellectual eminence, another peculiarity of *Homo sapiens*.

MERICARP. One-seeded portion of a schizocarp (q.v.).

MERISTELE. See *Dictyostele*.

MERISTEM. Localized region of active cell-division in plants from which permanent tissue is derived. New cells formed by activity of a meristem become variously modified to form characteristic tissues of the adult (epidermis, cortex, vascular tissue, etc.). A meristem may have its origin in a single cell, e.g. in ferns, or in a group of cells as in flowering plants. The principal meristems in the latter group occur at tips of stems and roots (apical meristems or growing points), between xylem and phloem of vascular bundles (cambium) in the cortex (cork cambium), in young leaves, and e.g., in many grasses, at bases of internodes (intercalary meristems). Meristems may also arise in response to wounding.

MEROBLASTIC. (Of animal zygote), only a part undergoing cleavage (q.v.). Occurs in yolky eggs (e.g. chick), the yolk-rich part remaining undivided into cells. Cf. *Holoblastic.*

MEROGAMETE. Of Protozoa, gamete formed by multiple division, and smaller than the ordinary organism, as distinct from hologamete which is as large as the ordinary organism and only one of which is produced from each individual. When male and female gametes are strongly differentiated (e.g. malaria parasite) male gametes are merogametes and female hologametes.

-MEROUS. As a suffix, referring to the number of parts, e.g. corolla pentamerous, consisting of five petals.

MESENCEPHALON. Mid-brain (q.v.).

MESENCHYME. Embryonic connective tissue consisting of rather widely scattered, irregularly branching cells in a jelly-like matrix. Gives rise to connective tissue, bone, cartilage, blood, etc.

MESENTERY. (1) Double layer of peritoneum attaching stomach, intestines, spleen, etc., to dorsal wall of peritoneal cavity. Contains blood, lymph, and nerve supply to these organs. (2) More strictly, the same fold attaching small intestine only. (3) Vertical partitions in coelenteron of Actinozoa.

MESODERM (MESOBLAST). Germ-layer of triploblastic animal embryo composed, like endoderm, of cells which have moved from surface of embryo into its interior during gastrulation; developing into tissues between gut and ectoderm (muscle, blood, connective tissue, etc.). Term is usually applied to the germ-layer while it is still a demarcated region of the embryo, after gastrulation but before differentiation into the derived tissues; and all the tissues at later stages derived from the embryonic layer may be called mesodermal. See *Germ layer*, *Ectoderm*, *Endoderm.*

MESOGLOEA. Layer of jelly-like material between external and internal cellular layers of the body of a coelenterate. It may be thin and contain prolongations of cells of these layers, as in *Hydra* and delicate hydroids; or bulky and tough (fibrous), containing many cells (which do not however form organs), as in sea-anemones or large jelly-fish.

MESONEPHROS. Part of kidney of vertebrates which arises in embryonic development later than, and posterior to, the pronephros (q.v.) and discharges into pronephric (Wolffian) duct. There is one on each side. The testes connect with them, and in adult sperms pass through them and down the Wolffian ducts to the exterior. Become functional kidneys of adult anamniotes. In amniotes they function as kidneys in embryo; but later in development are superseded in this function by metanephros (q.v.). and they degenerate except for part connected with testis in male. See *Epididymis*, *Vas efferens*, *Metanephros*.

MESOPHYLL. Internal tissue of leaf-blade differentiated into *palisade* and *spongy* mesophyll. In dorsi-ventral leaves palisade mesophyll consists of elongated cells immediately below upper epidermis with their long axes at right-angles to leaf surface; these cells contain numerous chloroplasts and are chiefly concerned in photosynthesis. Spongy mesophyll is a tissue of irregular, loosely arranged cells, containing fewer chloroplasts, with large air spaces between the cells, communicating through stomata with atmosphere outside.

MESOPHYTE. Plant growing under average conditions of water supply. Cf. *Hydrophyte*, *Xerophyte*.

MESOTHELIUM. Layer of flattened cells, a simple squamous epithelium (q.v.) in appearance, derived from mesoderm; lines coelomic cavities of vertebrates (see *Pericardium*, *Pleura*, *Peritoneum*), and some other spaces, e.g. synovial sacs.

MESOTHORAX. Middle of the three segments of the insect thorax. Bears a pair of walking-legs and (in winged insects) a pair of wings.

MESOZOIC. See *Geological Periods*. This era (the age of great reptiles) lasted approximately from 225 till 60 million years ago.

METABOLA. Pterygota (q.v.).

METABOLISM. (1) In general the chemical processes occurring within an organism, or within part of one. They involve breaking down of organic compounds from complex to simple (catabolism) with liberation of energy available for the organism's many activities; and building up of organic compounds from simple to complex (anabolism) using energy liberated by catabolism and, in the case of autotrophic organisms, energy from external non-organic sources (particularly from sunlight). Catabolism is by no means confined to the breaking down of the 'food-material' necessary for the organism's energy requirements. It involves, continuously, very many constituents of the organism, though they are broken down at various rates (DNA in resting cells is one of the constituents that seems to remain quite stable); and correspondingly anabolism involves, continuously, synthesis of many constituents. These processes go on even in an apparently static tissue like bone. This metabolic whirlpool, whose extent has become apparent only since the use of 'tracers' (q.v.) is a system of reactions predominantly controlled by enzymes (q.v.). In most of the aspects of metabolism which have been sufficiently investigated, such as catabolism of carbohydrates, there is a fundamental similarity throughout a wide variety of organisms, including plants, animals, and bacteria. Correspondingly, there are close similarities between the enzymes of these organisms. See *Metabolite*, *Respiration*. (2) Metabolism of some constituent of an organism (e.g. 'fat metabolism'); the part of total metabolism involving that constituent.

METABOLISM, BASAL. The rate of energy expenditure of an animal at rest, usually expressed per unit weight. In man, basal metabolic rate (B M R) is expressed as the output of Calories per square metre of body surface per hour. Measured directly, or indirectly by calculation from the amount of oxygen consumed or carbon dioxide given off.

METABOLITE. Substance which takes part in a process of metabolism (q.v.). Most metabolites are made by the organism in the course of metabolism; others must be taken in from the environment, the organism being unable to make them itself; in still other cases part of the supply of a particular metabolite may be made by the organism, part taken in. Autotrophic organisms need only inorganic metabolites, e.g. water, carbon dioxide, nitrates, and a number of trace elements (q.v.) from the environment. Heterotrophic organisms need, as well as inorganic metabolites, a wide range of organic metabolites from the environment, varying much from species to species, but including certain amino-acids and vitamins.

METACARPAL BONES. Bones of the fore-foot of tetrapod vertebrates (palm-region of man) Rod-like bones, usually one corresponding to each digit. Articulate with wrist bones (carpals) proximally, finger-bones (phalanges) distally. See Fig. 7, p. 175.

METACARPUS. Region of fore-foot of tetrapods, containing metacarpal bones. Palm-region of man.

METACHROMATIC. When stained with a dye, taking a colour different from that usually produced by the dye. E.g. heparin stains purple with thionin, instead of the usual blue. Such staining with basic dyes is commonly an indication of the presence of sulphated polysaccharides.

METACHRONAL RHYTHM. Pattern of beating of cilia (as in ciliated epithelium) or multiple limbs (as in Polychaeta). If the effective stroke is defined as occurring backwards, then each cilium or limb is at a slightly earlier stage in the beat cycle than the one behind it, slightly later than the one in front of it. The effective beat therefore appears to pass as a wave *forwards* in the series of cilia or limbs, though the surrounding fluid is driven *backwards*.

METAMERIC SEGMENTATION (METAMERISM). See *Segmentation*.

METAMORPHOSIS. Period of rapid transformation from larval to adult form. Often involves considerable destruction of larval tissues. E.g. transition of frog tadpole to adult.

METANEPHROS. Part of kidney of amniote vertebrates which arises in embryonic development later than, and posterior to, mesonephros (q.v.), one on each side. Acquires special duct, ureter, which grows into it from the original pronephric duct. Functional kidney of late embryo and adult. See *Pronephros*, *Renal Portal System*.

METAPHASE. Stage of mitosis (q.v.) of meiosis (q.v.) when chromosomes are arranged on equator of spindle. See Figs. 6A, 6B, pp 147, 148.

METAPLASIA. Transformation of one sort of normal adult tissue into another.

METATARSAL BONES. Bones in the hind-foot of tetrapod vertebrates. Rod-like bones, usually one corresponding to each digit. Articulate with ankle bone (tarsals) proximally, toe-bones (phalanges) distally See Fig. 7, p. 175.

METATARSUS. Region of hind-foot of tetrapods containing metatarsal bones.

METATHERIA. Marsupialia (q.v.).

METATHORAX. Hindmost of the three segments of insect thorax. Bears a pair of walking legs and (in many winged insects) a pair of wings.

METAXENIA. Influence of male (pollen) parent on maternal tissue (i.e. outside embryo and endosperm), e.g. variation in time of maturity of dates with different pollen parents. This particular effect is thought to be brought about by hormones secreted by developing embryo and endosperm.

METAXYLEM. Elements of primary xylem differentiated from procambium (q.v.), after protoxylem; wider, with thicker, more heavily lignified walls; inextensible.

METAZOA. Animals whose bodies consist of many cells, as distinct from Protozoa, which are unicellular; all animals commonly recognized as animals, including man. Sponges (Parazoa, q.v.) though also multicellular, differ so much from other multicellular animals that they are not usually included in the Metazoa.

MICROBE. Microscopic organism, usually pathogenic.

MICRODISSECTION. Technique of doing operations on minute objects (e.g. living cells) viewed through a microscope, by mechanically operated instruments.

MICROGAMETE. Male gamete, differentiated from female gamete by smaller size and/or by structure. See *Anisogamy*.

MICRO-INCINERATION. Method of examining distribution of minerals in sections of tissues, or whole micro-organisms, by placing on a slide, burning away all organic constituents in a high-temperature furnace, and determining the nature and position of the ash microscopically.

MICRON. Unit of length for microscopic objects. One thousandth of a millimetre. Usually written μ. The limit of resolution of a microscope using visible light is about one-fifth of a micron. A millimicron, one thousandth of a micron, is written $m\mu$, and is ten Angstroms (the physicists' unit of small distance).

MICRONUCLEUS. See *Macronucleus*.

MICRO-NUTRIENT. Substance which an organism must obtain from its environment to maintain health, though necessary only in minute amounts; either vitamin (q.v.) or trace element (q.v.).

MICRO-ORGANISM. Microscopically small organism; unicellular plant, animal, or bacterium.

MICROPHAGOUS. Of animals, feeding on particles minute relative to their own size. E.g. barnacle, right whale. Such animals tend to feed continually; and particularly by sieving particles from water drawn to them by ciliary action, or by movements of appendages. Plankton is main source of such particles; these feeders are all aquatic. Cf. *Macrophagous*.

MICROPYLE. (Bot.). Canal formed by extension of integument(s) of ovule beyond apex of nucellus; recognizable in mature seed as a minute pore in seed coat through which water enters when seed begins to germinate. (Zool.). Pore in egg-membranes of an ovum, through which sperm reaches and fertilizes the ovum, e.g. in many insects.

MICROSCOPE. The ordinary microscope of the laboratory is a *compound microscope* with two sets of lenses (objective and eye-piece) which magnify the object in two steps. Biologists work with various magnifications, achieved by using objectives of different powers, commonly a 'low-power' or 'two-thirds' objective (giving a final magnification including that of the eye-piece of roughly 60–100 times), a 'high-power' or 'sixth' objective (final magnification of roughly 200-400 times) and an oil immersion (q.v.) 'twelfth' objective (final magnification 600–1000 times, sometimes up to 1,500 times). 'Two-thirds', 'sixth' and 'twelfth' indicate the focal length of the objective in inches. About 1,500 times is the maximum degree of magnification it is useful to have using ordinary light, because it is sufficient to make it easy to look at the finest detail that can possibly be seen ('resolved'). Further magnification reveals nothing more; the details merely look bigger and vaguer. This is because it is physically impossible with a microscope, however great its magnification, to see that two points are really two and not a single point when they are nearer together than approximately half the wavelength of the light used to illuminate them. Detail finer than this cannot be resolved. A modern compound microscope can almost reach this physical limit. Only by decreasing the wave-length of the illumination used can finer detail be resolved; and that is how the electron microscope achieves its immense power of detecting minute detail, the 'illumination' being by electrons, which have a very short wavelength.

MICROSPORANGIUM. Sporangium within which microspores are formed. In flowering plants, pollen sac. Cf. *Megasporangium*.

MICROSPORE. Smaller of the two kinds of spore produced by heterosporous ferns; first cell of male gametophyte generation of these plants and of seed plants. In flowering plants, becomes pollen grain. Cf. *Megaspore*.

MICROSPOROPHYLL. Leaf or modified leaf bearing microsporangia. In flowering plants the stamen. Cf. *Megasporophyll*.

MICROTOME. Machine for cutting extremely thin slices of tissue (usually 3 to 20 microns; for the electron microscope, with special techniques, 100 times thinner than this). Such slices (*sections*) are easily stained and examined with a microscope. Tissue for cutting is either frozen, or, more usually, embedded in a firm but easily-cut supporting substance, generally paraffin wax.

MID-BRAIN (MESENCEPHALON). Middle of the three divisions marked out by constrictions in the embryonic vertebrate brain. Becomes during development very thick-walled, with small central cavity. Particularly concerned with sight and hearing.

MIDDLE EAR. See *Ear, middle.*

MIDDLE LAMELLA. Primary layer of wall of a plant cell on which secondary layers are later deposited; composed of pectic substances. See *Cell-wall.*

MILDEW. (1) Plant disease caused by a fungus that produces a superficial, powdery, or downy growth on surface of host; (2) Fungus causing such a disease; (3) often used synonymously with mould (q.v.).

MILK TEETH. Deciduous teeth (q.v.).

MILLON'S TEST. Test for phenols, including the amino acid tyrosine which is present in most proteins.

MIMETIC. Mimicking another species, see *Mimicry.*

MIMICRY. Protective similarity in appearance of one species of animal to another. In *Batesian mimicry* one of the two is poisonous, distasteful, or otherwise protected from predators, and often conspicuously marked (warning or aposematic coloration); the other, innocuous, gaining protection from predators by similarity to the first. In *Müllerian mimicry* both species are protected from predators, and they gain mutually from having the same warning coloration since predators can learn to avoid both species by tasting one. Mimicry occurs particularly amongst insects.

MIOCENE. Geological period (q.v.), sub-division of Tertiary, lasted approximately from 25 till 10 million years ago.

MIRACIDIUM. Ciliated larva of fluke (q.v.). Emerges from egg, which is released from vertebrate host in excreta, and parasitizes a snail, in which it reproduces asexually.

MITES. Members of the order Acarina other than ticks (q.v.). Some mites are free-living (e.g. cheese mite) others parasitic (e.g. *Sarcoptes*), causing scabies and mange.

MITOCHONDRIA (CHONDRIOSOMES). Minute semi-solid bodies numbers of which occur in the cytoplasm of every cell (except Bacteria and blue-green Algae). Enclosed in a membrane and have a complex internal structure of folded membranes. Rod-shaped or granular (name mitochondria sometimes restricted to latter); approximately $\frac{1}{4}$–1 micron width or diameter. Stain specifically in the living cell with Janus Green. Made of protein and fat and contain many enzymes, notably oxidative enzyme systems. Mode of multiplication unknown. Plastids of green plants are possibly modified mitochondria. See Fig. 1, p. 44.

MITOGENETIC RAYS. Short wave-length (ultra-violet) rays, supposed to emanate from living tissues and to stimulate mitosis in other tissues exposed to them. Their existence is widely doubted.

MITOSIS (KARYOKINESIS). The usual process by which a nucleus divides into two. Each chromosome duplicates, probably before beginning of mitosis, and mitosis involves separation of resulting duplicates so that one goes into each daughter nucleus. As a result the two daughter nuclei have an identical complement of chromosomes, and hence of genes. The process is divided into four stages: prophase, metaphase, anaphase, telophase (Fig. 6A). *Prophase* (2) Within the optically homogeneous resting nucleus chromosomes appear, at first as very long threads (the whole tangle is sometimes called a spireme) which shorten and thicken steadily by each coiling into a close spiral. Each chromosome is probably already double at its earliest appearance. *Metaphase* (3). Nuclear membrane dissolves. A spindle (q.v.) forms where the nucleus was, between the centrioles (q.v.) if these are present. The chromosomes lie (as an

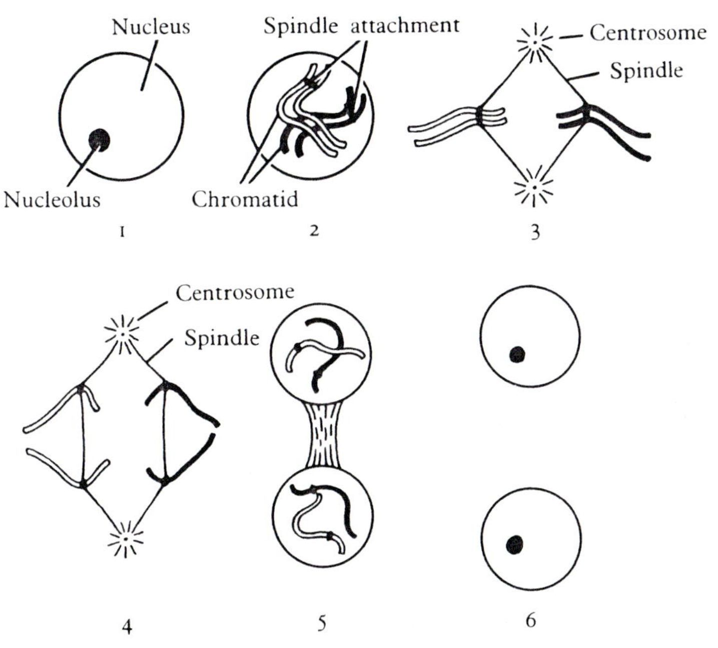

Fig. 6A. Diagram of mitosis in a nucleus showing only one pair of chromosomes (one chromosome black, other white). There are usually many pairs in each nucleus. Cytoplasm is not shown. Note that centrosomes do not occur in cells of most plants. 1, *Resting stage.* 2, *Prophase, each chromosome already double (consisting of two chromatids).* 3, *Metaphase.* 4, *Anaphase.* 5, *Telophase.* 6, *The two daughter nuclei in resting stage.*

'equatorial plate') at the equator of the spindle, attached to it by their spindle attachments (q.v.). They pause in this position. *Anaphase* (4). The duplicates of each chromosome (chromatids) separate and move rapidly towards the poles of the spindle. They do not reach the poles, but the spindle itself elongates, pushing the two groups of chromosomes further apart. *Telophase* (5). The chromosomes uncoil, elongating and finally disappearing. A new nuclear membrane forms. Spindle gradually disappears. The cytoplasm, if it divides, does so at this stage (see *Cell-division*).

Time taken for mitosis varies a good deal but is usually between half and three hours. See *Meiosis, Amitosis.*

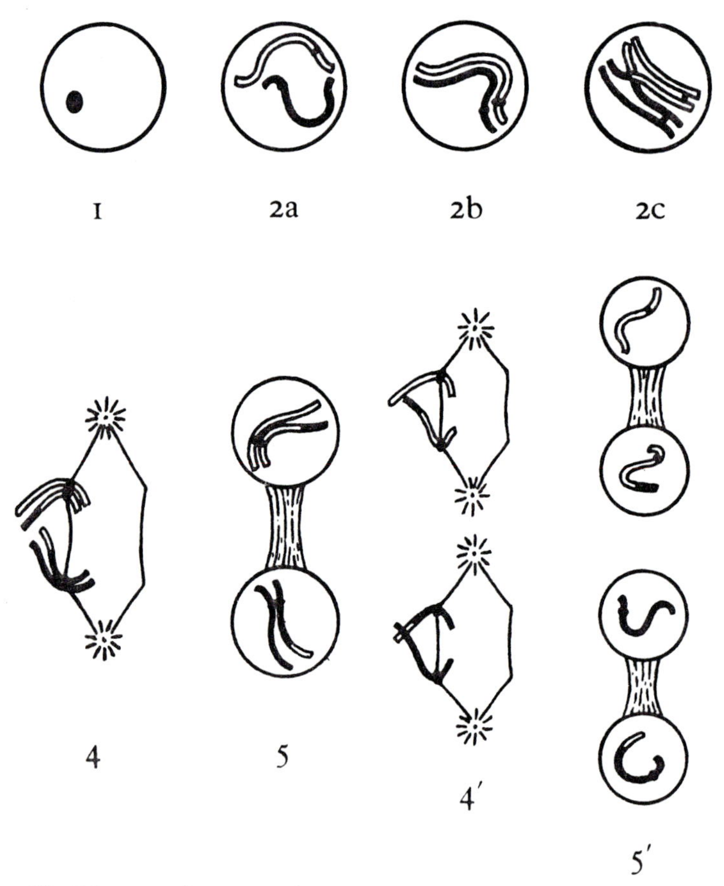

Fig. 6B. Diagram of certain stages of meiosis, in nucleus like that of fig. 6A. Stages are numbered to correspond to those of mitosis shown in fig 6A Prophase in three stages, 2a (appearance of single *chromosomes), 2b (pairing), 2c (doubling of chromosomes with chiasma formation). 4 and 5, first anaphase and telophase, two chromatids having mixed constitution due to chiasma. 4′ and 5′, second anaphase and telophase. Four nuclei result from the whole process of meiosis.*

MITRAL VALVE. Valve between left auricle and left ventricle of mammalian heart, consisting of two membranous flaps.

MOLARS. Those crushing back teeth of mammals which, unlike premolars, have no predecessors in the milk teeth. They have usually several roots, and a complicated pattern of ridges and projections on biting surface. See *Dental Formula*.

MOLLUSCA. Large phylum of animals including mussels, snails, octopuses, etc., mostly aquatic; soft bodied, often with a hard shell; unsegmented; with a head and muscular foot. Includes classes Amphineura, Gastropoda, Scaphopoda, Lamellibranchiata, Cephalopoda.

MOLLUSCOIDA. Term formerly used for a heterogeneous assemblage of animals supposed to resemble the Mollusca, e.g. Brachiopoda, Urochordata, Polyzoa, Phoronidea.

MONADELPHOUS. (Of stamens), united by their filaments to form a tube surrounding the style, e.g. lupin, hollyhock. Cf. *Diadelphous*, *Polyadelphous*.

MONOCARPIC. (Of a plant), flowering once during its life.

MONOCHASIUM. Kind of inflorescence (q.v.).

MONOCHLAMYDEOUS (HAPLOCHLAMYDEOUS). (Of flowers), having only one whorl of perianth segments.

MONOCOTYLEDONEAE. Smaller of the two classes into which flowering plants (Angiospermae) are divided; distinguished from the larger class, *Dicotyledoneae*, by presence of a single seed leaf (cotyledon) in the embryo and by other structural features, e.g. parallel veined leaves, stem vascular tissue in form of scattered closed vascular bundles, flower parts usually in threes or in multiples of three. A few monocotyledons are large plants, e.g. palm trees, but majority are small. Group includes many important food plants, e.g. cereals, fodder grasses, bananas, palms; and ornamentals, e.g. lily, tulip, orchid.

MONOCYTE. Largest kind of white blood cell (q.v.) of vertebrate blood (9–12 microns diameter in man), with spherical nucleus. Form 3–8 per cent of white blood cells in man. Actively phagocytic when, in inflammation, they invade the tissues. Identical with or closely related to the macrophages of the tissues.

MONOECIOUS. (1) (Of plants). Having both male and female reproductive organs on the same individual; in flowering plants having unisexual, male and female, flowers on same plant e.g. hazel. Cf. *Dioecious*, *Hermaphrodite*. (2) (Of animals) hermaphrodite (q.v.). Cf. *Dioecious*, *Unisexual*.

MONOKARYON. (Of fungus hypha or mycelium) composed of cells each containing one haploid nucleus; characteristic of phase of life-cycle of many Basidiomycete fungi. Cf. *Dikaryon*.

MONOPHYLETIC. (Of a taxon) consisting of individuals descended from a common ancestor which is a member of the same taxon. Cf. *Polyphyletic*.

MONOPODIUM. Axis produced and increasing in length by apical growth, e.g. trunk of pine tree. Cf. *Sympodium*.

MONOSACCHARIDE. Simplest sugar; ceases to have the properties of a sugar when split up. See *Hexose*, *Pentose*, *Disaccharide*.

MONOTREMATA. A sub-class of Mammalia, containing the duck-billed platypus and two genera of spiny anteater. Live in Australia and New Guinea. A very primitive group, only distantly related to all other known mammals: from which they differ in laying eggs, and in possessing many other reptilian features. But have hair and secrete milk. Fossil history unknown.

MONOZYGOTIC TWINS (IDENTICAL TWINS, UNIOVULAR TWINS). Twins due to fission into two, at some stage of its development, of the embryo derived from a single fertilized egg. Such twins are genetically identical and therefore of same sex. Cf. *Dizygotic twins*, *Polyembryony*.

MORPHOGENESIS. Development of form or structure in ontogeny or in regeneration.

MORPHOGENETIC MOVEMENTS. Displacements of masses of cells in embryonic development of animals, particularly conspicuous during gastrulation (q.v.).

MORPHOLOGY. Study of form.

MORULA. (Zool.). Embryo during process of cleavage, before blastula stage; consists of a number of blastomeres.

MOSAIC. (1) (Bot.). Symptom of many virus diseases of plants, a patchy variation of normal green colour. (2) (Genetics). Synonym of chimaera (q.v.). Animals of this kind (i.e. a mixture of cells of different genetic composition), such as gynandromorphs, are more usually called mosaics, plants usually chimaeras. (3) (Zool.). *Mosaic development*; development when determination is complete but before functional differentiation (q.v.) starts, each region of the embryo differentiating almost independently of influence from other regions. Little regulation (q.v.) can occur. Time of onset of mosaic development varies much. In vertebrates does not start till after gastrulation; in many invertebrates (e.g. groups with spiral cleavage) zygotes or early cleavage stages show mosaic development, and hence are known as *mosaic eggs*.

MOSSES. Musci (q.v.).

MOTOR. Concerned with stimulation of effector organs. *M. cortex*. Part of cerebral cortex controlling nerves to striped muscles. *M. end-plate*. End-organ (q.v.) of a motor nerve-fibre on a muscle fibre, consisting of an accumulation of muscle cytoplasm and nuclei, within which the nerve-fibre branches, and through which its nerve-impulses stimulate the muscle fibre. *M. nerve*. Peripheral nerve consisting of nerve-fibres of motor neurones. *M. nerve-fibre*. Nerve-fibre of motor neurone. *M. neurone* (motoneurone). Nerve-cell whose nerve-fibre connects with effector organ; conducts impulses from central nervous system (q.v.) which stimulate effector to activity. See *Nerve-root*.

MOTOR ROOT. Nerve-root, synonymous with central root (q.v.).

MOULD. (1) Any superficial growth of fungus mycelium (mildew); (2) popular name for fungus.

MOUTH-PARTS. Structures which surround the mouth of arthropods and are concerned in feeding. They are mostly modified paired appendages of the head segments (see *Labrum*, *Mandible*, *Maxilla*, *Labium*).

MUCIN. Mucoprotein forming mucus (q.v.) in solution.

MUCOPROTEINS (GLYCOPROTEINS). Compounds of polysaccharide with protein; e.g. mucin, matrix of cartilage.

MUCOSA. Mucous membrane (q.v.).

MUCOUS MEMBRANE. General name for a moist epithelium (q.v.) and its immediately underlying connective tissue, in vertebrates. Particularly applied to lining of gut and urino-genital ducts. Epithelium is usually simple, though stratified near openings to exterior; often ciliated; often contains goblet cells (q.v.) secreting mucous.

MUCUS. (1) Slimy solution of mucin, secreted by goblet cells of vertebrate mucous membrane. (2) Any viscous, sticky or slimy fluid (usually of unknown composition) secreted by invertebrates.

MÜLLERIAN DUCT. Oviduct of female gnathostome vertebrate. A pair in most species, only one in birds. A tube, at one end opening into coelom by a ciliated funnel, at other end joining cloaca (or, with urethra, joining remnant of cloaca in placental mammals). Muscular and ciliary movements pass eggs down the tube; and where fertilization is internal, pass sperm up it. In marsupial and placental mammals it is differentiated into fallopian tube, uterus, and vagina (but q.v.). In all placentals the posterior parts of the pair of ducts fuse, so that there is a single median vagina, and often, as in man, a single median uterus. The duct has no representative, or only a vestige, in males. Embryonically arises from mesothelium of coelom in close association with Wolffian duct (q.v.); it is therefore mesodermal.

MULTICELLULAR. Consisting of many cells. Cf. *Unicellular*.

MULTINUCLEATE. Of a cell (q.v.), containing many nuclei.

MULTIPLE ALLELOMORPHS. A set of more than two genes, all allelomorphs (q.v.). Only two of the set can be present at the same time in somatic cells of an animal or a vascular plant (except in polyploids, q.v.). Multiple allelomorphs originate from each other by mutation.

MUSCI. Mosses, a class of green plants grouped with the Hepaticae (Liverworts) in division Bryophyta. World-wide in distribution, occurring in damp habitats, e.g. moist woodland, in water, and under drier conditions, e.g. on heaths, walls. A moss plant consists of a prostrate or erect stem, a fraction of an inch to a foot in length, bearing closely arranged leaves and anchored to the substratum by rhizoids. Internally the stem possesses a distinct central conducting strand of elongated cells. Sexual organs (antheridia and archegonia) are borne in separate groups at tips of stems. Male gametes are motile by flagella. Fertilization is followed by development of a capsule, containing spores, each of which, on germination, gives rise to a filamentous or thalloid *protonema*, from which new moss plants develop as lateral buds. Differ from liverworts in their more complex organization of plant body and sporophyte.

MUSCLE. Tissue consisting of cells which are highly contractile. Most of edible parts of animals (flesh) is muscle. See *Cardiac muscle*, *Actomyosin*, *Smooth muscle*, *Striped muscle*. MUSCLE FIBRE see *Striped muscle*.

MUSCULO-EPITHELIAL CELL. Cell characteristic of coelenterates, forming greater part of body of a simple member of the group such as *Hydra*. Columnar cell, arranged with others in a continuous sheet, each cell with one or two processes which are contractile, and which extend into the mesogloea. The main parts of the cells form the epithelial outer or inner layers of the body. In some coelenterates (jelly-fish) epithelial part may be reduced so that the cell looks more like a smooth muscle cell.

MUSHROOM. Popular name for edible fruit-bodies of Fungi belonging to family Agaricaceae of Basidiomycetes, especially of species of *Psalliota*.

MUTANT. Gene which has undergone mutation within the particular stock of organisms under observation; or organism bearing such a gene; or character due to such a gene.

MUTATION. Sudden and relatively permanent chromosomal change. The most important mutations are those occurring in the gametes or their precursors, since they can produce an inherited change in the characteristics of the organisms developing from them, such changes being the raw material of evolution. The majority of mutations are changes of individual genes (gene-mutations), but some are gross structural alterations of chromosomes (e.g. inversion, translocation) or changes in numbers of whole chromosomes per nucleus (e.g. polyploidy). Mutation is normally a very infrequent event, though it can be greatly speeded up by irradiation with X-rays, gamma-rays, neutrons, etc., and by some chemicals (e.g. mustard gas). From the standpoint of adaptation of the organism, mutations are random. Evolution has occurred by natural selection of mutations, not by directed mutation. They are merely the raw material for evolutionary change. The great majority of mutations are deleterious, upsetting the balanced mechanism of embryonic development of the organism; and on the whole the larger the change they induce the more deleterious they are.

Mutation in the restricted sense of gene-mutation is a sudden change in an individual gene so as to alter its effects on the cellular processes it influences. The changed gene is allelomorphic to the old. It reproduces itself, in the changed form, many times (perhaps of the order of a million) before undergoing another mutation, which may be back to the original gene, or to a new allele. The same kind of gene mutation, i.e. a change from one particular allele to another particular allele, occurs repeatedly within a population of organisms, probably over long periods of time. E.g. in the human population the gene for haemophilia (q.v.) has arisen repeatedly by mutation from a normal allele. An entirely new gene-mutation is therefore a much rarer event than the mutation rate would suggest.

A mutation may occur in a body-cell (*somatic mutation*) and is then transmitted to all cells derived by mitosis from that cell.

MYCELIUM. Collective term for mass of hyphae that constitutes vegetative part of a fungus, e.g. mushroom spawn.

MYCETOZOA. Myxomycetes (q.v.).

MYCOLOGY. Study of Fungi.

MYCORRHIZA. 'Fungus root'; an association of a fungus with root of a higher plant. Mycorrhizas are of common occurrence. Two main types, (1) *endotrophic*, in which fungus is within cortex cells of root, e.g. orchid, and (2) *ectotrophic*, in which it is external, forming a mantle that completely invests the smaller roots, e.g. pine tree. Mycorrhizas were formerly believed to constitute an example of a symbiotic association of mutual benefit. More recently it has been suggested that they are merely examples of limited parasitic attacks. In some cases at least the benefits from the association, e.g. in the establishment of seedling trees such as pine, presence of mycorrhizal fungus has been shown to be vital.

MYCOSIS. Disease of animals caused by fungal infection, e.g. ringworm.

MYCOTROPHIC. (Of plants). having mycorrhizas.

MYELIN. Fatty substance (phospholipid, cholesterol, etc.) with protein, making a sheath to larger nerve fibres (q.v.) of vertebrates and Crustacea.

MYELOID TISSUE. Tissue producing myeloid elements (polymorphs and red blood cells) of vertebrate blood, which are formed throughout life. Except in embryo, usually located in bone marrow. In adult man mainly in ribs, sternum, skull, hip girdle, vertebrae.

MYOGLOBIN. A variety of haemoglobin (q.v.) occurring in muscle fibres.

MYONEME. Contractile fibril of Protozoa.

MYOSIN. See *Actomyosin*.

MYOTOME. That part of somitic mesoderm (q.v.) which forms striped muscle.

MYRIAPODA. Centipedes and millipedes. Class of Arthropoda with long bodies and many legs, distinct head, one pair of antennae, mandibles, at least one pair of maxillae; live on land and breathe air by means of tracheae. Centipedes (Chilopoda) have one pair of legs per segment; are usually more or less flattened; carnivorous. Millipedes (Diplopoda) have two pairs of legs per segment (each segment representing two embryonic segments fused together); are usually cylindrical; and herbivorous. There are several important differences between these two groups which indicate that they are not very closely related, and they are sometimes put into separate classes.

MYXAMOEBA. Naked cell, capable of amoeboid movement, characteristic of vegetative phase of Myxomycetes and certain simply organized Fungi.

MYXOMYCETES (MYCETOZA). Slime moulds. Group of very simple organisms with both plant and animal characteristics; taxonomic position uncertain, usually included in division Thallophyta. In the vegetative condition they consist of naked, multinucleate masses of protoplasm known as *plasmodia.* These show amoeboid movement and ingest food (animal characteristics). They reproduce by spores, shown in some genera to have cellulose walls, formed within sporangia (a plant characteristic). Meiosis is presumed to occur before spore formation. The spores germinate to form one or more amoeboid, biflagellate swarm cells which may either behave as gametes and copulate in pairs soon after their formation, or may first lose their flagella, undergo a series of divisions and then copulate. Growth of the zygote results in formation of a plasmodium. Plasmodia may also be formed by coalescence of many zygotes or small plasmodia. Widely distributed, occurring under damp conditions on decaying vegetable matter. One of the commonest species, *Fuligo septica* (flowers of tan), forms yellow plasmodia up to eight inches in diameter and frequently occurs on tanner's bark.

MYXOPHYCEAE. Cyanophyceae (q.v.).

N

NARES. Nostrils of vertebrates. Opening from olfactory organ: *external nares* on to surface of head, or *internal nares* into mouth. See *Choanae.* Usually paired.

NASAL CAVITY. Cavity of tetrapod head containing olfactory organ communicating with mouth and with surface of head by internal and external nares respectively; lined by mucous membrane. See *Palate.*

NASTIC MOVEMENT. (In plants) response to a stimulus that is independent of direction of stimulus. May be a growth curvature, e.g. opening and closing of many flowers in response to changes in light intensity; or a sudden change in turgidity of particular cells causing a rapid change in position of particular plant organs, e.g. rapid folding of leaflets and drooping of leaves of sensitive plant (*Mimosa pudica*) when lightly struck. Nastic movements are classified according to nature of stimulus, e.g. *photonasty*, response to alteration in light intensity, *thermonasty*, to alteration in heat intensity, *seismonasty*, to shock. Cf. *Tropism*.

NATIVE PROTEIN. Protein in its naturally occurring stage, i.e. not denatured. See *Denaturation*.

NATURAL ORDER. See *Family*.

NATURAL SELECTION. The principal mechanism of evolutionary change, originally suggested by Darwin in 1859. The theory that evolution occurs by natural selection asserts that, of the range of different individuals which make up the population of a given species, those individuals having certain characteristics contribute more offspring to the succeeding generation than those having other characteristics; and if such characteristics have an inherited basis, the composition of the population is thereby changed.

If one considers the zygotes produced within a population of a species, usually only a minority will develop to maturity and still fewer will succeed in producing offspring. Within such a population there is always variation between individuals. The theory of natural selection asserts that the contribution of offspring to the next generation is not entirely random, but is correlated with this variability. Some kinds of variant individual are consistently more successful in leaving offspring than other kinds. The successful variants and their progeny are said to be 'selected', by a natural process. In other words different variations (q.v.) have different 'survival value' in the face of the hostile circumstances in which all organisms live. Of these variations, only those which are inherited are important for evolution. Natural selection is thus the main agent controlling the composition of a population during the course of time, eliminating certain variations and thus preventing change in some directions, making other variations more prevalent and hence producing evolutionary change in other directions.

NATURE AND NURTURE. Synonymous with genotype and environment respectively. Differences in either or both of these may be responsible for differences in character (variation, q.v.) between organisms. Variations due to 'nurture' are not usually inherited.

NAUPLIUS. Kind of larva of many crustacean species. Oval, unsegmented, three pairs of appendages.

NEANDERTHAL MAN. *Homo neanderthalensis*, a rather recently (perhaps 50,000 years ago) extinct species of hominid from Pleistocene (mainly from early part of fourth glaciation about 100,000 years ago). Human brain size, but heavy brow ridges, low forehead, no chin prominence, did not walk fully erect. Its remains are associated with more than one culture, but particularly Mousterian, which had ritual burial of dead.

NECTARY. Gland secreting sugary fluid (*nectar*) attractive to insects in many insect pollinated flowers.

NEKTON. Swimming animals of pelagic zone of sea or lake. Includes, e.g. fishes and whales. Cf. *Plankton.*

NEMATOCYST. See *Thread-cell.*

NEMATODA. Round, thread-, eel-worms. Phylum of animals not closely related to any other, with many peculiarities of, e.g. muscular, excretory and reproductive systems, body cavity, and development. Includes minute free-living forms, e.g. vinegar eels; plant parasites, e.g. eelworm of potatoes; and animal parasites, e.g. hook-worm.

NEMERTEA. Ribbon worms; a small phylum of marine worms sharing several characteristics with the Platyhelminthes, but differing from them in having a tube-like gut with mouth and anus, a peculiar proboscis, a simpler reproductive system and a circulatory system.

NEO-DARWINISM. Name sometimes given to modern evolution theory which combines the theory of natural selection with the discoveries of Mendelian genetics.

NEO-LAMARCKISM. See Lamarkism.

NEOLITHIC. Phase of human history during which plants were cultivated and animals domesticated, and many other technical advances on previous (palaeolithic) period made. Started approximately 10,000 years ago in Middle East and spread slowly over the world (though there was probably independent discovery of cultivation in America).

NEOPALLIUM. That part of roof of cerebral hemispheres of vertebrates which is not particularly connected with sense of smell, but serves more general co-ordination. Forms main mass of cerebral cortex of man.

NEOPLASM. Tumour. Abnormal localized multiplication of some type of cell. *Malignant* if the growing cells infiltrate surrounding tissue and are carried by blood or lymph to other localities in the body, there continuing their growth; otherwise *benign.*

NEOTENY. Persistence of the form of a larval or of other early stage of development. Either temporary, e.g. climatically delayed development; or permanent, in which case the animal must breed in juvenile form (paedogenesis, e.g. axolotl, and many other animals which as adults resemble juvenile stages of related species). Neoteny may be applied to retarded development of individual structures, e.g. ostrich feathers resembling down-feathers of flying birds. Neoteny has probably been very important in evolution of many groups, including man, who has resemblances to young stages of apes.

NEOTYPE. Specimen selected to replace the type specimen (q.v.)

when all the original material is lost or destroyed. Cf. *Lectotype.*

NEPHRIDIUM. Organ present in many invertebrates (Platyhelminthes, Nemertea, Rotifera, Annelida, larvae of some Mollusca) and in amphioxus. Probably excretes, and helps to control water content of body. Consists of a tube of ectodermal origin opening to the exterior at one end. The other end may be closed (*protonephridium*) with flame cells opening into it; or it may open into the coelom. A nephridium may combine with a coelomoduct (q.v.) forming a *nephromixium.*

NEPHRON. Excretory unit of vertebrate kidney, consisting of a Malpighian corpuscle (q.v.) and its attached uriniferous tubule (q.v.).

NERITIC. Inhabiting the sea over the continental shelf; arbitrarily taken to be sea where it is shallower than 200 metres. Cf. *Oceanic.*

NERVE. Bundle of motor and/or sensory nerve-fibres with accompanying connective tissue and blood-vessels, in a common sheath of connective tissue. Each nerve-fibre conducts impulses independently of its fellows. See *Cranial Nerve, Spinal Nerve.*

NERVE-CELL (NEURON, NEURONE). (1) Cell of nervous tissue which conducts the impulses (q.v.) by which the nervous system functions. Each has a nucleus surrounded by a mass of cytoplasm, constituting the cell-body or perikaryon; and projecting from this, thread-like processes, very various in length and number, which carry the impulses from place to place. A commonly occurring arrangement of these processes is a single long *axon* (q.v.) which carries impulses away from the cell-body to other nerve-cells or to effectors; and numerous short *dendrites* (q.v.) which receive impulses from the axons of other nerve-cells. Transfer of impulses from nerve-cell to nerve-cell takes place at synapses (q.v.). Central nervous system (q.v.) of all animals contains very numerous nerve cells. Some are entirely confined within it, forming an elaborate interconnecting system. Others are only partly within it, their thread-like processes running out of it to all parts of the body, some to effectors (motor nerve fibres), some to receptors (sensory nerve fibres). Other nerve cells are situated mainly or wholly outside the central nervous system, e.g. those in nerve nets (q.v.) or sensory nerve cells. (2) Term is sometimes used synonymously with cell-body (q.v.).

NERVE CORD. Solid strand of nervous tissue, forming part of central nervous system (q.v.) of invertebrates. In Annelida and Arthropoda there are two parallel ventral nerve cords, each bearing a row of segmentally arranged ganglia.

NERVE ENDING. The structure at the peripheral end of a fibre of the peripheral nervous system, in which impulses start (in a sensory fibre) or finish (in a motor fibre). May be merely the bunch of fine branching twigs by which the fibre itself ends (*free nerve ending*); or may be a distinct end-organ (q.v.).

NERVE-FIBRE. The axon of a nerve-cell (q.v.), or the similar process (but carrying impulses towards the cell body) of many peripheral sensory nerve cells, together with the various special membranes which may surround it. Many nerve-fibres of vertebrates and some of Crustacea and annelid worms are *myelinated* (*medullated*): i.e. the axon is covered by a relatively thick fatty sheath of myelin except at periodically-spaced nodes of Ranvier. Fibres which lack a myelin sheath are called *unmyelinated* or *non-medullated*; their axons are bounded merely by the usual plasma-membrane of all cells. Surrounding the axons of fibres of the vertebrate peripheral nervous system lie the Schwann cells, of which the myelin is really a part; and each single myelinated fibre is further enclosed in a reticulin neurilemmal tube. In the central nervous system Schwann cells and neurilemmal tubes are absent. The diameter of nerve fibres varies in any vertebrate animal from 1 to 20μ; in some invertebrates giant nerve fibres occur and may reach 1 mm. diameter in squids and burrowing worms. A nerve-fibre commonly branches into many small-diameter twigs at its termination. An increased diameter gives greater conduction velocity; the same velocity is also produced in smaller space by a myelin layer, the impulses then apparently leaping from node to node along a fibre instead of travelling continuously along it. A nerve may be nearly as long as the whole animal, so that in large vertebrates fibres may occur up to a million times longer than they are thick.

NERVE IMPULSE. See *Impulse*.

NERVE-NET. Network of nerve-cells, diffusely distributed through tissues, which in some phyla (e.g. Coelenterata, Echinodermata) makes up all or most of the nervous system, though usually associated with some concentration of the cells into major strands providing through ways for conduction. In phyla with well-developed central nervous systems (e.g. Arthropoda, Vertebrata, Cephalopoda) a nerve-net may hardly exist (usually only in gut wall). Consists of separate nerve-cells communicating with each other by contact only (synapses). Conduction in a nerve-net is slow and often in all directions, because the synapses are not, or are only weakly, one-way transmitters. In Coelenterata a strong stimulus, eliciting more impulses, has a more extensive effect than a weak one, because impulses are used up in producing summation at the synapses; not, as previously thought, because of any fading of impulses within the nerve-fibres.

NERVE ROOT. In vertebrates, each spinal nerve arises from the spinal cord by two roots; a *dorsal* (posterior in human anatomy) and a *ventral* (anterior in human anatomy) which join to form the spinal nerve as they pass through the wall of the vertebral column. The dorsal roots contain all the sensory nerve fibres, their cell-bodies (q.v.) being in a ganglion on the course of each root; in some vertebrates they may also contain a few motor fibres. Through the ventral roots pass the motor fibres, their cell-bodies being in the spinal cord. Cranial nerves can also be classified into dorsal and ventral roots, which do not however join, remaining separate nerves; cranial dorsal roots have ganglia and contain, besides sensory, numerous motor fibres running to face, jaw, and gill muscles, and motor fibres of the parasympathetic system; cranial ventral roots contain some sensory (proprioceptor, q.v.) fibres. Dorsal and ventral roots are a characteristic feature of all vertebrates, and occur in amphioxus, though in this chordate motor-fibres to the viscera go through dorsal roots, and (as also in some Cyclostomes) the two roots do not join to form spinal nerves.

NERVOUS SYSTEM. A mechanism which co-ordinates the various activities of an animal with each other and with events in the external world, by means of messages rapidly conducted from part to part. Present in all multicellular animals except sponges. It consists of numerous cells of special kind (nerve-cells, q.v.), with branching thread-like processes. Every nerve-cell is in contact with others by means of these processes, the points of contact being called synapses (q.v.). The nerve-cells and their processes, linked by synapses, form a system which permeates the whole body. Some of the processes of the nerve-cells terminate at sense-organs, some at muscles or other effectors (q.v.), while many end at synapses with other nerve-cells. The nervous system functions by 'messages' (impulses, q.v.), which run along the nerve-cell processes and are able to cross synapses. Impulses are started by sense-organs, and they can set a muscle or other effector in operation when they reach it. But the nervous system does not simply provide a means by which the stimulation of a certain sense-organ automatically activates a certain effector. That would limit the number of possible activities to the number of direct receptor-effector connexions. There usually are many fairly direct connexions of this type. See *Reflex*. But the work of the nervous system co-ordinates activities in a far more subtle and complicated way, by means of the synapses. A synapse discriminates which impulses it passes through to the succeeding nerve-cell according to what impulses are arriving at other synapses involving that nerve cell. See *Summation* and *Inhibition*. These other synapses link up with other parts of the nervous system. Consequently, an impulse started by a sense-organ is not limited to the two possibilities of activating or failing to activate a given receptor. It can, by means of connecting synapses, promote or inhibit numerous other activities, according to the state of other parts of the nervous system. Very complicated co-ordination is, therefore, possible with suitably complex connexions. In most animals, the elaborate co-ordination for which the synapses are an essential basis occurs in the central nervous system (q.v.). Here most of the synapses are located, and the rest of the nervous system (peripheral nervous system, q.v.) consists mainly of nerve-cell processes running directly to sense-organs or effectors. The nervous system is not the only co-ordinating system. There is a much more slowly acting and less flexible one based on hormones (q.v.).

NERVOUS TISSUE. Nerve cells (q.v.) and/or their nerve fibres; together with the accessory cells which closely surround them, e.g. Schwann cells of vertebrate nerve fibres; and supporting connective tissue or glia (q.v.) with blood-vessels.

NEURAL. Concerned with the nervous system. *N. arch*, a bony arch resting on centrum of each vertebra, forming tunnel (*N. canal*) through which spinal cord runs. *N. crest*, embryonic material of vertebrates, found initially at both sides of the neural plate as this rolls up and sinks beneath epidermis; gives rise to extraordinary variety of tissues, probably including dorsal root ganglia, Schwann cells, sympathetic ganglia, melanophores, cartilage of visceral arches. *N. fold*, raised ridge of neural plate, neural crest, and epidermis along edge of neural plate, beginning the formation of neural tube. *N. plate*, flat expanse of neural tissue, first-formed embryonic rudiment of nervous system of vertebrate. See *Organizer*. *N. spine*, median dorsal bony projection from neural arch, serving usually as muscle attachment. *N. tissue*, embryonic nervous tissue, in vertebrates a simple epithelium of columnar cells which only later differentiate into glia and nerve cells. *N. tube*, longitudinal tube of neural tissue in vertebrate embryos, formed by rolling up of neural plate until the neural folds join in the mid-dorsal line to form a tube and the epidermis fuses above; expanded in front as the rudiment of the brain, the narrower posterior part being that of spinal cord; forms central nervous system, and motor nerve fibres of peripheral nervous system.

NEUROCRANIUM. The part of the skull surrounding brain and internal ear, as distinct from the part composing jaws and their attachments (*splanchnocranium*)

NEUROGLIA. Glia (q.v.).

NEUROHUMORAL. Acting by transmission of the effects of a nerve impulse, from a nerve fibre to an end-organ or across a synapse, by secretion of minute amount of chemical substance from the nerve fibre See *Cholinergic*, *Adrenergic*.

NEURONE (NEURON). Nerve-cell (q.v.).

NEURONE THEORY. Theory that nervous system is made up of numerous discrete nerve cells, each originating from single cell of embryonic neural tissue, which merely make contact with each other at synapses; in contrast to the view (now largely discarded) that there is protoplasmic continuity throughout the nervous system.

NEUROPTERA. Order of endopterygote insects including alderflies, lacewings. Two similar pairs of membranous wings which when at rest are held up over body. Larvae of some are aquatic, others, e.g.

lacewing larvae, feed on aphides and help to control these pests.

NEURULA. Stage of vertebrate embryo after most of the gastrulation movements have ceased, when determination is far advanced and differentiation, manifested externally by neural plate, is proceeding fast. Stage ends when neural tube complete.

NEUTER. Sexless. Without reproductive organs but with other parts more or less normal.

NICOTINIC ACID (NIACIN, P-P FACTOR). A vitamin of the B group, lack of which is part of the cause of pellagra in man. Forms part of a respiratory coenzyme, widely (universally?) distributed. Required as a vitamin by those vertebrates and insects tested. Synthesized by many micro-organisms.

NICTITATING MEMBRANE. Transparent fold of skin forming a third eye-lid. When open, lies at inner (anterior) corner of eye or below lower eye-lid. Occurs in some sharks and amphibia, and widespread in reptiles and birds. Well-developed in few mammals.

NIDICOLOUS BIRDS. Those which hatch in relatively undeveloped state and stay in nest some time after hatching.

NIDIFUGOUS BIRDS. Those which hatch well-developed, and leave the nest immediately.

NIT. Egg of human louse, which is cemented on to hair.

NITRIFICATION. Conversion by soil bacteria of organic compounds of nitrogen, unavailable to green plants, into available nitrates. Occurs in several steps, involving different kinds of bacteria. See *Nitrogen Cycle*.

NITROGEN CYCLE. Circulation of nitrogen atoms brought about mainly by living things. In outline, inorganic nitrogen compounds (chiefly nitrates), are absorbed from soil or sea and built up into organic compounds by autotrophic plants. These die or decay or are eaten by animals and the nitrogen, still as organic compounds, returns to soil or sea in their excreta or by their death and decay. Bacteria then convert them to inorganic nitrogen compounds. Some nitrogen is lost to the atmosphere as gas, and some is extracted from the atmosphere by nitrogen-fixing (q.v.) bacteria and blue-green algae.

NITROGEN FIXATION. Conversion of atmospheric nitrogen into organic nitrogen compounds, a process that can be carried out only by certain soil-inhabiting bacteria and certain blue-green algae (*Cyanophyceae*). Some nitrogen-fixing bacteria live symbiotically with leguminous plants (peas, beans, clover, etc.) in nodules on roots, others live independently in soil. By their activity soil is enriched in nitrogen, a fact of considerable practical importance. See *Nitrogen Cycle*.

NODE. (1) Part of plant stem where one or more leaves arise. Cf. *Internode*. (2) Nodes of Ranvier. Constrictions regularly repeated along myelinated peripheral nerve fibres.

NOMENCLATURE. See *Binomial Nomenclature*.

NOTOCHORD. Skeletal rod (of large vacuolated cells packed within a firm sheath) lying lengthwise, between C.N.S. and gut. Present in some stage of development of all chordates. In most vertebrates occurs complete only in embryo (or larva), though remnants usually persist, between the vertebrae which replace it, in the adult. Occurs in larval and adult amphioxus, in larval tunicates, and (it is often held) is represented in adult Hemichordata.

NUCELLUS. Central tissue of ovule (q.v.), containing embryo sac and surrounded by integument(s).

NUCLEIC ACID. Long chain compound formed from a large number of nucleotides; found in all living things. *Deoxyribosenucleic acid* (DNA) is formed from nucleotides each of which contains the sugar deoxyribose and usually one of four different bases (2 pyrimidine, 2 purine). It is now widely thought to be the inherited material (see *Gene*), and as such to have a great variety of structure, produced by different arrangements of the nucleotides with different bases; it is thought to be able to pass on this structure to copies of itself (self-replication), and the structure is thought to be translated into the complex structure of proteins when the latter are synthesized. It is found almost exclusively in the chromosomes of plants and animals and in the corresponding structures of bacteria and viruses. *Ribosenucleic acid* (RNA) is formed from nucleotides containing the sugar ribose and, as with DNA, usually one of four bases. One of these bases is different from those occurring in DNA (uracil instead of thymine). RNA is the material of inheritance of some viruses, and in all living things seems to take an essential part in protein synthesis, as an intermediary between the structure of DNA and that of protein.

NUCLEO-CYTOPLASMIC RATIO. Ratio of volume of nucleus to volume of cytoplasm; may be fairly constant for any given cell-type of a given species.

NUCLEOLUS. Small dense body containing ribose nucleo-protein (q.v.) one or more of which occurs inside the resting nucleus and are visible during life. Disappears during mitosis; produced by certain regions of chromosomes. See Fig. 1, p. 44.

NUCLEO-PROTEINS. Compounds of nucleic acid and protein.

NUCLEOTIDE. Compound formed from one molecule of each of a sugar (with 5 carbon atoms), of phosphoric acid, and of a nitrogen-containing base (purine or pyrimidine). Found free in cells (e.g. ATP, q.v.) and as part of various coenzymes, and as the building-blocks of nucleic acids (q.v.).

NUCLEUS. Body containing the chromosomes (q.v.) present in nearly all cells of plants and animals and probably in bacteria; not in viruses. Every nucleus has originated from a previous nucleus, usually by mitosis (q.v.) or meiosis (q.v.), occasionally by amitosis (q.v.). Variously shaped, usually spherical or ovoid with firm superficial membrane. In the non-dividing cell it usually appears homogeneous in living state, except for one or more nucleoli; and when fixed contains a darkly staining (basophilic) irregular meshwork. It is highly probable that the organization of the chromosomes is intact in resting cells, and that the homogeneity is deceptive and the meshwork an artefact. At onset of mitosis or meiosis, chromosomes become visible in both living and fixed cell, separating out from a non-basophilic *nuclear sap*.

Nucleus is essential for continued life of most cells (but see *Red Blood Cell*); if it is removed (as by microdissection) remaining cytoplasm soon dies. See Fig. 1, p. 44.

NUCLEUS (OF BRAIN). Demarcated mass of nerve cell-bodies, i.e. of grey matter, in vertebrate brain. A very large number of such nuclei have been anatomically distinguished, though their function is often uncertain. Nuclei are connected by tracts of nerve-fibres. The term is anatomical, not physiological. Cf. *Centre*.

NUNATAK. See *Refugium*.

NUT. Dry, indehiscent, one-seeded fruit, somewhat similar to an achene but product of more than one carpel and usually larger with a hard, woody wall, e.g. hazel nut.

NUTATION (CIRCUMNUTATION). Spiral course pursued by apex of a plant organ during growth due to continuous change in position of most rapidly growing region of the organ; most pronounced in stems but also occurs in tendrils, roots, flower-stalks, sporangiophores of some fungi.

NYCTINASTY. Response by plants to periodic alternation of day and night, e.g. opening and closing of many flowers, 'sleep-movements' of leaves. Related to changes in temperature, light intensity. See *Photonasty*, *Thermonasty*.

NYMPH. Young stage of exopterygote insect; resembling the adult in, e.g. kind of mouth parts, and in having compound eyes (cf. *Larva*); different in being sexually immature, and wingless or having wings incompletely developed.

O

OBDIPLOSTEMONOUS. (Of stamens), in two whorls, outer opposite petals (instead of alternating with petals, the general condition), inner opposite sepals, e.g. stitchwort.

OCCIPITAL CONDYLE. Bony knob at back of skull articulating with first vertebra. See *Condyle*, *Atlas*. Absent in most fish, whose skull is not movable on vertebral column. Double in Amphibia and mammals, single in reptiles and birds.

OCCIPUT (OCCIPITAL REGION). (1) In vertebrates, an inexactly delimited region of the head in neighbourhood of joint between skull and vertebral column. (2) In insects, plate of exoskeleton forming back of the head.

OCEANIC. Inhabiting sea where it is deeper than 200 metres. Cf. *Neritic.*

OCELLUS. Simple light-receptor occurring in many invertebrates.

OCTOPODA. Suborder of Mollusca (class Cephalopoda); octopus, argonaut.

OCULOMOTOR NERVE. Third cranial nerve of vertebrates. Almost entirely motor, supplying four of the extrinsic eye-muscles (q.v.; the inferior oblique muscle, and all rectus muscles except external); and, by nerve fibres of parasympathetic system, intrinsic eye-muscles of accommodation and pupil constriction. A ventral root.

ODONATA. Dragon-flies. Order of exopterygote insects with aquatic nymphs. Carnivorous; have large eyes; two pairs of similar wings. Strong fliers. Some fossil dragon flies (Carboniferous) had a wing span of two feet.

ODONTOBLASTS. Cells which lie in pulp cavity of vertebrate tooth and send processes into adjacent dentine (q.v.) which they take part in forming.

ODONTOID PROCESS. See *Atlas.*

OEDEMA (EDEMA). Swelling of tissue through increase of its (chiefly intercellular) fluid content, due to passage of extra amounts of water out from capillaries, e.g. in inflammation. Also used of generalized overdevelopment of plant cells.

OESOPHAGUS. Part of gut between pharynx and stomach, concerned simply in passing food along by peristalsis.

OESTROGEN. Any substance producing changes in genital tract characteristic of follicular phase of oestrous cycle (q.v.) in various test mammals. Oestrogenic activity is usually tested on mouse vagina (q.v.). An oestrogen (probably oestradiol) is normally secreted by mammalian ovary and is responsible for many of the phenomena of the oestrous cycle (q.v.) and for development and maintenance of many female sexual characteristics. Also produced by placenta during pregnancy, and to small extent by adrenal cortex and testis of mammals. Probably secreted by ovaries of all vertebrates.

OESTROUS CYCLE. Reproductive cycle of short duration (usually between five and sixty days according to species) occurring in sexually mature females of many species of mammal, in absence of pregnancy (and only during breeding season, where such occurs). Each cycle consists of a brief period (usually a day or two) of oestrous or 'heat' at which time and at no other the female will copulate with a male, ovulation (q.v.) coinciding with this oestrus; and, preceding and succeeding oestrus, there are various changes throughout the body, particularly in the uterus, which may be regarded as preparations for pregnancy. The whole set of changes is controlled by hormones. In outline the phases of the cycle are typically (*a*) *follicular* with growth of Graafian follicles (q.v.) in ovary, proliferation of lining of uterus (endometrium), increasing secretion of oestrogen (q.v.) by ovary; (*b*) *ovulation*, with activation of mating reflexes; (*c*) *luteal*, with corpus luteum (q.v.) formation in ovary, great development of uterine glands, secretion of progesterone (q.v.) by corpora lutea, oestrogen secretion diminishing; (*d*) regression of corpus luteum, beginning of new follicular growth, return to unproliferated state of uterine lining, diminution of oestrogen and cessation of progesterone secretion. If fertilization occurs in phase (*b*) then phase (*d*) is omitted, and the cycle is suspended in the luteal phase for the duration of pregnancy. If there is no fertilization then the cycle usually immediately repeats, but in some species there is only one cycle per breeding season. Cycle depends on cyclical production of gonadotrophic hormones (q.v.) by pituitary gland, these hormones controlling changes in the ovary which in turn controls all the other changes by means of the cycles of oestrogen and progesterone production. There is considerable variation according to species in the characteristics of the cycle; the luteal phase may for instance be omitted (mouse); and another variant is the menstrual cycle (q.v.).

OIL-IMMERSION OBJECTIVE. Objective of light microscope (q.v.), space between which and the cover-slip over the object examined is filled with drop of oil of same refractive index as glass; system used for finest resolution and highest magnification with the light microscope.

OLECRANON PROCESS. Bony process on ulna of mammals extending beyond elbow joint, for attachment of muscles which straighten fore-limb.

OLFACTORY. (Adj.). Of sense of smell. *O. bulb*, terminal part of cerebral hemisphere of vertebrates, from which springs *O. nerve* (first cranial nerve) running to organs of smell.

OLIGOCENE. Geological period (q.v.), subdivision of Tertiary, lasted approximately from 40 till 25 million years ago.

OLIGOCHAETA. Order of annelids (class Chaetopoda) including earthworms. Freshwater and terrestrial; chaetae few; no parapodia; head without special appendages; hermaphrodite; fertilization internal; clitellum present; eggs develop in a cocoon. Cf. *Polychaeta*.

OMMATIDIUM. See *Eye, Compound*.

OMNIVOROUS. Eating a diet of both plants and animals. Cf. *Carnivore*, *Herbivore*.

ONCHOSPHERE. (Hexacanth.) Six-hooked embryo of tapeworms. Develops from egg, and if it gets into suitable host, grows into larva (cysticercus, cysticercoid, or plerocercoid).

ONTOGENY. The whole course of development during an individual's life history.

ONYCHOPHORA. Small group of animals all belonging to one genus, *Peripatus*. Worm-like, with many short unjointed legs. Found in warmer parts of the world. With some arthropod characters, e.g. tracheae, but have thin unjointed cuticle and other characters which they share with annelids. Sometimes classified as a sub-phylum or class of Arthopoda, sometimes as a separate phylum.

OOCYTE. (Zool). Cell which undergoes meiosis and thereby forms ovum. *Primary oocyte* undergoes greater part of cytoplasmic growth involved in ovum formation, and first of the two meiotic divisions. As a result of this division it gives rise to one cell with very little cytoplasm (polar body, q.v.) and one *secondary oocyte* with massive cytoplasm. Latter undergoes second meiotic division, giving rise to another polar body and an ovum. Fertilization frequently occurs in one of the two oocyte stages. Cf. *Spermatocyte*.

OOGAMY. Sexual reproduction in which a large, non-motile egg is fertilized by a small male gamete. Occurs in all Metazoa and some plants. See *Anisogamy*.

OOGENESIS. (Zool.). Formation of ova. See *Maturation of Germ-Cells*, *Oocyte*.

OOGONIUM. (1) (Bot.). Female sex organ of certain Algae and Fungi, containing one or more oospheres. (2) (Zool.). Cell of animal ovary which undergoes repeated mitosis and eventually gives rise to oocytes (q.v.).

OOSPHERE. (Bot.). Large, naked, spherical, non-motile female gamete (egg); formed within an oogonium.

OOSPORE. (Bot.). Thick-walled resting spore formed from a fertilized oosphere.

OPERCULUM. (1) Lid, e.g. of moss capsule, egg shell of fluke, (2) Cover of gill-slits of fish and Amphibia. (3) Exoskeletal plate of some gastropods, e.g. snail, which can close opening of shell when animal withdraws into the latter.

OPHIDIA. Snakes. A sub-order of Squamata, or sometimes made a separate order of Reptilia. Limbless; with exceptionally wide jaw gape due to mobility of bones; eyelids immovable, nictitating membrane fused over cornea; no ear drums.

OPHIUROIDEA. Brittle-stars; class of Echinodermata; star-shaped; long and sinuous arms radiating from clearly delineated central disc; easily break up by autotomy.

OPHTHALMIC. Optic (q.v.).

OPTIC. (Adj.). Of the eye. *O. chiasma*, structure formed beneath vertebrate forebrain by nerve-fibres of right optic nerve crossing to left side of brain and vice versa. In most vertebrates all the nerve-fibres cross. In mammals, however, 50 per cent remain on their original side, 50 per cent cross. *O. nerve*, second cranial nerve of vertebrate, really part of brain wall. See *Retina*.

ORAL. (Adj.). Of the mouth.

ORBIT. Cavity or depression in skull of vertebrates housing eyeball.

ORDER. One of the kinds of group used in classifying organisms. Consists of a number of similar families (sometimes of only one family).

Similar orders are grouped into a class. 'Natural orders' of flowering plants are equivalent to Families (q.v.). See *Classification*.

ORDOVICIAN. Geological period (q.v.), lasted approximately from 500 till 440 million years ago.

ORGAN. Part of an animal or plant which forms a structural and functional unit, e.g. leaf, kidney.

ORGANELLE. Of Protista, a specialised part of the cell, roughly analogous to *organ* of Metazoa, e.g. flagellum, contractile vacuole.

ORGANIZER. (1) Dorsal lip of blastopore of amphibian embryo, consisting of material which will become notochord and somites; and corresponding region of other vertebrate embryos. When transplanted to blastula of same or related species, performs a complex of inductions (q.v.) as a result of which the transplanted organizer and the surrounding host tissues develop into a complete embryonic axis; the best-known amongst these inductions is the evocation (q.v.) of neural plate. (2) Any part of an embryo which performs an induction on another part.

ORGANOGENESIS. Differentiation of organs.

ORNITHINE CYCLE. Probable method of urea formation in ureotelic (q.v.) vertebrates. Carbon dioxide and ammonia combine with the amino acid ornithine to give the amino acid arginine, which is split by enzyme arginase into ornithine and urea. Occurs in liver. See *Uricotelic*.

ORNITHISCHIA. Fossil order of Reptilia, formerly part of order Dinosauria which is now divided into Ornithischia and Saurischia. Lived during Mesozoic. Contained only herbivorous species and none reaching the gigantic size of the biggest Saurischia.

ORNITHORHYNCHUS. The Duckbilled Platypus, a very primitive monotreme mammal of Australia.

ORTHOGENESIS. Steady trend of evolution in a given direction over a prolonged period of time, affecting related groups of organisms, due to the working out of inherent trends within the inherited material. It is not generally accepted that orthogenesis is the proper explanation of the many known instances of persistent evolutionary trends.

ORTHOPTERA. Order of exopterygote insects including cockroaches, locusts, grasshoppers. Mostly large, terrestrial insects with biting mouth parts; narrow, hardened forewings, and membranous hind wings. Good runners and jumpers, many flightless.

ORTHOTROPOUS. (Of ovule), erect on funicle, micropyle pointing away from placenta. Cf. *Anatropous*, *Campylotropous*.

OSCULUM. Of sponges, an opening through which the water taken in through ostia leaves the body.

OSMIUM TETROXIDE (OSMIC ACID). OsO_4. Component of all the best fixatives for cytological purposes, since it produces little distortion. Blackens fat.

OSMO-REGULATION. Control of osmotic pressure within an organism. See *Contractile vacuole*, *Internal environment*.

OSMOSIS. When a solution of, say, sugar in water is separated from pure water by a membrane permeable to water but not to sugar (*semi-permeable membrane*), water passes across the membrane into the sugar solution. This movement of water is *osmosis*. If external pressure is applied to the sugar solution, the movement of water into it will be opposed. The pressure required to stop the movement completely is called the *osmotic pressure* of the sugar solution. It is greater the more concentrated the solution. Water will similarly pass by osmosis from any solution having a weaker osmotic pressure to any having a stronger (involving the same or different dissolved substance) provided the membrane separating them is of the appropriate semi-permeable kind, until the two solutions attain equal osmotic pressures. And if the membrane is merely *more* permeable to water than to the dissolved substance, water will still move from weak solution to strong, while the dissolved substance moves more slowly in the opposite direction, until the two solutions have equal osmotic pressures. The amount of osmotic pressure produced by solutions (in standard conditions) depends on the *number* of dissolved particles (molecules or ions), whatever their size, in a given amount of water. Proteins, and other substances with large molecules, develop therefore a small osmotic pressure relative to the weight of them which is dissolved. More or less semi-permeable membranes are very widespread in living organisms, e.g. they surround all cells; and so osmosis plays a great part in controlling the distribution of water in living organisms.

OSTEICHTHYES. Bony fish, often treated as a single class of vertebrata, though by others divided into two classes, Actinopterygii and Choanichthyes (q.v.).

OSTEOBLAST. Cell responsible for formation of calcified intercellular substance of bone. Cells within bone (*osteocytes*) are osteoblasts which have become included as the bone developed.

OSTEOCLAST. Multinucleate cell which breaks down the calcified intercellular substance of bone. Remodelling of bone shape by such breaking down constantly accompanies bone growth.

OSTIOLE. Pore in fruit bodies (pycnidia and perithecia) of certain Fungi and in conceptacles of brown Algae through which, respectively, spores or gametes are discharged.

OSTIUM. Of sponges, an opening through which water is drawn into the body. Much more numerous than oscula (q.v.).

OSTRACODERMI. Fossil Agnatha. A grouping not now much used in classification: it is split up into separate orders. All have well-developed bony armour. Ordovician to Devonian.

OTIC. Concerning the ear. See *Auditory*.

OTOCYST. Statocyst (q.v.).

OTOLITH. Granule of calcium carbonate in vertebrate inner ear.

Several such granules are attached to fine processes of sensitive cells, the latter communicating via nerves with the brain. The pull of gravity on the granules and therefore on the cell-processes registers the position of the animal with respect to gravity. Similar arrangement in some invertebrates. See *Statocyst.*

OUTBREEDING. See *Inbreeding.*

OVARIAN FOLLICLE. Sac of cells which invests developing oocyte of many Metazoa. Probably concerned in nourishment of growing oocyte, and in vertebrates secretes oestrogen. See *Graafian Follicle.*

OVARY. (1) (Bot.). Hollow basal region of a carpel, containing one or more ovules. In a flower which possesses two or more united carpels the ovaries are united to form a single compound ovary. (2) (Zool.). Organ which produces ova. In vertebrates it also produces sex hormones. See *Oestrogen, Progesterone.*

OVIDUCT. (Zool.). Tube carrying ova from ovary, or from coelom into which ova are shed, to the exterior. See *Müllerian Duct.*

OVIPAROUS. Laying eggs in which the embryos have as yet developed little if at all. Cf. *Ovoviviparous, Viviparous.*

OVIPOSITOR. Organ at hind end of abdomen in female insects, through which eggs are laid. Formed from modified parts of paired appendages; consists of several separate but interlocked parts. Frequently long and sometimes capable of piercing animals or plants, permitting eggs to be laid in otherwise inaccessible places, such as inside other organisms. Sting of bees and wasps is a modified ovipositor.

OVOGENESIS. Oogenesis (q.v.).

OVOTESTIS. Organ of some hermaphrodite animals, e.g. snail, which functions both as ovary (q.v.) and testis (q.v.).

OVOVIVIPAROUS. Having embryos which develop within the maternal organism, from which they may derive nutriment, though they are separated from it by the persistence, through most or all of development, of egg-membranes. E.g. many insects, snails, fish, lizards, and snakes. Cf. *Oviparous, Viviparous.*

OVULATION. Bursting of ripe egg from ovarian follicle. In vertebrates the egg (actually a secondary oocyte) is discharged on to surface of ovary and thence passes into oviduct; and ovulation is frequently result of stimulation by pituitary hormone. See *Oestrous Cycle.*

OVULE. Structure found in seed plants (gymnosperms and angiosperms) that develops into a seed after fertilization of an egg-cell within it. In gymnosperms the ovules are unprotected; in angiosperms protected by the megasporophyll which forms a closed structure (carpel) within which they are formed, singly or in numbers. Each ovule is attached to the carpel wall by a stalk or funicle which arises from its base (chalaza). A mature angiosperm ovule consists of a central mass of tissue, the nucellus, surrounded by one or two protective layers, the integuments, from which the seed coat is ultimately formed. Integuments enclose nucellus except at apex, where there is a small passage, the micropyle. Within the nucellus is a large oval cell, the embryo sac (q.v.) developed from megaspore, containing the naked egg-cell.

OVUM. Unfertilized egg-cell. A single large immobile cell, containing a haploid nucleus. In many animals, it is the oocyte (q.v.) which is fertilized by the sperm and in these animals, strictly speaking, an unfertilized ovum does not therefore exist.

OXIDASE. Enzyme (a kind of dehydrogenase, q.v.) which catalyses oxidation of a substrate by removal of hydrogen which combines with molecular oxygen.

OXYGEN DEBT. Oxygen consumed, in excess of normal amounts, when an organism or part of an organism has been respiring with inadequate oxygen supply; e.g. after hard muscular work oxygen consumption of man remains above normal for some time, until the debt has been repaid.

OXYHAEMOGLOBIN. Oxygenated haemoglobin. See *Respiratory Pigment.*

OXYTOCIN. Hormone secreted by posterior lobe of pituitary which produces strong contraction of uterine muscle. A polypeptide.

P

P_1. Parental generation, consisting of the parents from which a breeding experiment starts; all the males being uniform as regards the allelomorphs under observation, and all the females likewise.

PACEMAKER. Region of vertebrate heart where contraction at each beat is started: the sinus venosus (a contractile chamber of heart); or its homologue in mammals and birds the sino-auricular node (a small group of muscle and nerve cells in auricle wall). A wave of electro-negativity spreads at each beat from pacemaker first to auricle(s) then to ventricle(s), activiating them to contract in turn. If pacemaker removed, rest of heart beats, but at slower rate. Heart muscle has therefore intrinsic power of rhythmical contraction, normally masked by more rapid pacemaker stimulation.

PAEDOGENESIS. Reproduction in larval or other pre-adult form. See *Neoteny.*

PAIRING (OF CHROMOSOMES (SYNAPSIS). Side by side association of homologous chromosomes at meiosis (q.v.). See Fig. 6B, p. 148.

PALAEOBOTANY. Study of fossil plants.

PALAEOCENE. Geological period (q.v.), subdivision of Tertiary, not universally used, equivalent to early Eocene.

PALAEOLITHIC. Phase of human history during which, though tools were manufactured, food was obtained solely by hunting, fishing, or collecting, with no cultivation. Lasted from at least half a million years ago up to beginning of Neolithic stage (q.v.) about 10,000 years ago.

PALAEONTOLOGY. Study of fossils.

PALAEOZOIC. Geological era lasting approximately from 600 till 225 million years ago. See *Geological Periods*.

PALATE. Roof of vertebrate mouth. In mammals and crocodiles roof of mouth is not homologous with that of other vertebrates; a new (false) palate has developed beneath original palate, by bony shelves projecting inwards from bones of upper jaw. The nasal cavity is thus extended and its internal opening (choana, q.v.) placed right back in the throat. In mammals the bony part of the false palate or *hard palate* is continued backwards by a fold of mucous membrane and connective tissue, the *soft palate*.

PALEA (SUPERIOR PALEA, PALE). Glume-like bract of grass spikelet on axis of individual flower which, with lemma, it encloses.

PALISADE. See *Mesophyll*.

PALPS. Of polychaete annelid worms, tactile appendages on the head; of bivalve molluscs, ciliated flaps of tissue around mouth concerned in production of feeding currents; of Crustacea, distal parts of appendages which carry mandibles, may be locomotor, or may help in feeding; of insects, parts of first and second maxillae, shown to be olfactory in some insects.

PALYNOLOGY. Pollen analysis (q.v.).

PANCREAS. Sweetbread. Gland of gnathostome vertebrate situated in mesentery near duodenum, into which it discharges through pancreatic duct an alkaline mixture of digestive enzymes (trypsinogen, lipase, amylase, maltase, etc.). Stimulated to secrete this mixture by hormone secretin (q.v.). Also contains groups of cells (islets of Langerhans) of quite different function, secreting into blood a hormone, insulin (q.v.). Components of the pancreas probably occur in cyclostomes, but not united into a definite organ.

PANCREATIN. Extract of pancreas containing digestive enzymes.

PANICLE. Kind of inflorescence (q.v.).

PANTOTHENIC ACID. Vitamin of B group, forming a co-enzyme; required by a variety of organisms, e.g. some yeasts, some bacteria, insects, vertebrates.

PAPAIN. Intracellular proteolytic enzyme (or mixture of enzymes), found in a plant (paw-paw), which splits proteins in neutral solution.

PAPILLA. A projection. *Dermal papillae*, projections of dermis into epidermis of vertebrates, increasing surface of contact between the two tissues. *Tongue papillae*, variety of projections on upper surface of mammalian tongue, on which lie taste-buds, and which may be cornified to form a rasping surface, e.g. in cat.

PAPPUS. Ring of fine, sometimes feathery hairs, developed from calyx crowning the fruits of flowering plants of family Compositae, e.g.

dandelion; acting as a parachute and aiding in wind dispersal of fruits.

PARABIOTIC TWINS. Artificial 'siamese twins' produced by surgically joining two animals, so that their blood circulations become continuous.

PARAFFIN SECTIONS. Sections of tissues cut by microtome (q.v.) after embedding in paraffin wax. Usual method of preparing tissues for microscopical study.

PARAMECIUM. Genus of ciliate Protozoa. 'Slipper-animalcule.'

PARAMORPH. Any taxonomic variant within a species; used particularly when, because of lack of data, its status cannot be defined more precisely.

PARAPHYSIS. (Bot.). Sterile filament, numbers of which occur in Mosses and in certain Algae, interspersed amongst the sex organs; and in the hymenium of ascomycete and basidiomycete Fungi.

PARAPODIUM. Paired, segmentally arranged, muscular lateral projection of body of polychaete worms, bearing chaetae and sometimes other structures; locomotor in free-living forms.

PARASEXUAL CYCLE. Novel and flexible breeding system recently demonstrated in certain fungi (including both those with normal sexual reproduction and those with only asexual reproduction) that involves genetic recombination outside the sexual cycle. System comprises the following steps: (1) rare (1 in 10^6 to 10^7) fusion of two, genetically unlike nuclei in a heterokaryotic mycelium to produce a diploid heterozygous nucleus which (2) multiplies by mitosis during which crossing over occasionally occurs followed, again at rare intervals (1 in 10^3) by (3) formation of haploid nuclei from the diploids (haploidization) in which whole chromosomes reassort at random producing new genetic combinations. Starting from a culture containing a mixture of genetically different haploid nuclei the result of the parasexual cycle is a much more varied mixture of nuclei consisting of haploids like the originals, haploid recombinants, homozygous and heterozygous diploids. Little is yet known of natural significance of the parasexual cycle but it has already been shown to determine variation in pathogenicity in certain plant pathogens.

PARASITE. Organism living in or on another organism (its host) from which it obtains food. *Facultative p.*, an organism which can live entirely as a saprophyte (q.v.) but can live as a parasite under certain conditions. *Obligate p.*, an organism which can live only parasitically. Parasites may or may not be harmful to the host. Cf. *Symbiosis*, *Saprophyte*.

PARASYMPATHETIC SYSTEM. Part of autonomic nervous system (q.v.).

PARATHYROID. Endocrine gland of tetrapod vertebrates, usually paired, lying near or, in some species, within thyroid. Developed from gill pouches. Its hormone (a protein) controls distribution of calcium and phosphate in the body. Administration of hormone transfers calcium from bones to blood; deficiency lowers blood calcium, producing tetany.

PARATONIC. (Of plant movements), induced by external stimuli, e.g. nastic movements, taxes and tropisms (q.v.). Alternatively described as induced or aitiogenic. Cf. *Autonomic.*

PARAZOA. Grade of organization exemplified only by Porifera (q.v.), which are multicellular animals separately evolved from all others (Metazoa, q.v.).

PARENCHYMA. (Bot.). Tissue consisting of living, thin-walled cells, often almost as broad as long, and permeated by a system of intercellular spaces containing air. Cortex and pith are typically composed of parenchyma. (Zool.). Loose tissue consisting of irregularly-shaped vacuolated cells, forming a large part of body of Platyhelminthes. (2) The specific tissue of an organ, as opposed to the blood-vessels, connective tissue, etc.

PARIETAL. (1) Of coelomic lining (peritoneum or pleura), that covering inside of body-wall, as opposed to *visceral* part covering organs within coelom. (2) *P. bones*, pair of membrane-bones which in most vertebrates covers large part of upper surface of brain, behind frontal bones. (3) *P. organ*, see *Pineal apparatus*. (4) (Bot.). See *Placentation.*

PARTHENOCARPY. Formation of fruit without fertilization. In most plants development of a fruit from the ovary can proceed only after fertilization has occurred, but in a number of plants fruit development regularly takes place from unfertilized flowers. Such fruits, e.g. banana, pineapple, are seedless but otherwise normal in appearance. Parthenocarpy can also be induced in unfertilized flowers, e.g. in tomato, by application of certain hormones (q.v.). It has been suggested that fertilization provides a stimulus for production of hormone which, in turn, induces development of fruit tissue, and that where natural parthenocarpy occurs, sufficient hormone to enable fruit development to proceed is produced by the plant independently of stimulus of fertilization.

PARTHENOGENESIS. Development of ovum without fertilization (q.v.) into a new individual. In many animals it may be induced artificially. See *Artificial Parthenogenesis.* In some plants (dandelion), and animals (aphids, rotifers, where males may be absent), is of normal occurrence. Ova which develop in this way are usually diploid, in which case all offspring are genetically identical with the parent. Frequently in such animals ordinary sexual reproduction, providing genetic recombination, together with employment for males, occurs from time to time (heterogamy, q.v.). See *Apomixis.*

PASSAGE CELLS. Cells of endodermis, characteristically of older monocotyledonous roots, opposite protoxylem groups of stele, that remain unthickened, with casparian strips (q.v.) only, after thickening (involving deposition of suberin, cellulose, and lignification) of all other endodermis cells; allowing transfer of material between cortex and vascular cylinder.

PASSERINE. A member of the Passeriformes, a group of birds. A perching bird, with large first toe directed back, other three toes forward. The group includes about half the known species of bird, including most of our common inland ones.

PATELLA. Knee-cap. A bone (sesamoid bone, q.v.) over the front of the knee joint in tendon of (extensor) muscles which straighten the hind-leg, present in most mammals, some birds and reptiles.

PATHOGEN. Parasite which causes disease.

PATHOLOGY. Study of disease or of diseased tissues.

PEAT. Accumulated dead plant material that has remained incompletely decomposed owing, principally, to lack of oxygen, as in moorland and fen where land is more or less completely waterlogged; often forming a layer many feet deep.

PECTIC COMPOUNDS. Acid polysaccharide carbohydrates present in cell walls of unlignified tissue; comprising pectic acid and pectates, pectose (protopectin), pectin. Form gels under certain conditions. Principal constituents are galacturonic acid, galactose, arabinose, and methyl alcohol.

PECTORAL FIN. See *Fin*.

PECTORAL GIRDLE. Shoulder girdle (q.v.).

PEDICEL. Stalk of an individual flower of an inflorescence.

PEDIPALP. Second head appendage of arachnids; may be locomotor (kingcrabs); clawed and used for seizing prey (scorpions); sensory; or modified in the male for fertilization (spiders).

PEDUNCLE. Stalk of an inflorescence.

PELAGIC. Inhabiting the mass of water of sea or lake, in contrast to the sea or lake bottom. See *Benthos*. Pelagic animals and plants are divided into plankton and nekton (q.v.).

PELECYPODA. Lamellibranchiata (q.v.).

PELVIC FIN. See *Fin*.

PELVIC GIRDLE. Hip-girdle (q.v.).

PELVIS. (1) The pelvic girdle. (2) The lower part of the abdomen surrounded by the pelvic girdle. (3) *P. of kidney*, funnel-shaped expansion of ureter as it joins the concave side of kidney.

PENETRANCE. Proportion of organisms bearing a particular dominant gene, or homozygous for a recessive, which show the effect of that gene. Penetrance of many genes is practically 100 per cent but in the case of other genes it is much less, the value being affected by environment or genotype.

PENIS. Organ associated with duct from testis, used to introduce sperm from male into female.

PENTADACTYL LIMB. The kind of limb found in four classes of vertebrates: Amphibia, reptiles, birds, and mammals, collectively called the tetrapods. Evolved as an adaptation to life on land, and therefore not found in fishes or other primitively aquatic vertebrates. The limb is in three parts: upper-arm or thigh; fore-arm or shank; hand or foot. The latter bears five terminal fingers or toes (digits), hence the name pentadactyl. The first part contains one long bone (humerus in arm, femur in leg); the second two long more or less parallel bones (radius and ulna in arm, tibia and fibula in leg); the third many small bones in a fairly uniform pattern (carpals, metacarpals, and phalanges successively in arm; tarsals,

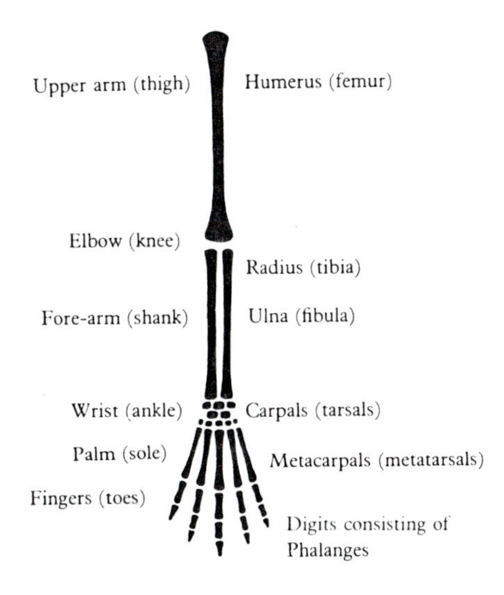

Fig. 7. Diagram of skeleton of pentadactyl limb of vertebrate, giving names of parts of fore-limb, and (in brackets) of hind-limb. On left, common name of whole part; on right, names of bones.

metatarsals, and phalanges in leg). See Fig. 7, above. Many modifications of this fundamental pattern occur, through loss or fusion of elements, especially in the terminal parts.

PENTOSE. Sugar (monosaccharide) with five carbon atoms, e.g. ribose. Important constituents of nucleic acids (q.v.); make up various plant polysaccharides, e.g. pectin, gum arabic.

PEPSIN. Enzyme (peptidase, q.v.) splitting proteins in acid solution. Secreted in vertebrate stomach, along with hydrochloric acid.

PEPTIDASE. An enzyme splitting peptides, and in many cases proteins, by attacking certain peptide links. *Exopeptidases* split off terminal amino-acids of a peptide chain. *Endopeptidases* split peptide chains by attacking links between certain amino-acids anywhere in the chain; e.g. trypsin, pepsin, papain; since they break up proteins into large fragments they are often called *proteinases*.

PEPTIDE. Compound formed of two or (polypeptide) more amino acids, by amino (NH_2) group of one joining with carboxyl

(COOH) group of the next, forming *peptide link* (– NH – CO –) with elimination of water.

PEPTONE. Product of protein splitting; more complex than peptides (q.v.), but not sharply demarcated from them.

PERENNATION. (Of plants) survival from year to year by vegetative means.

PERENNIAL. Plant that continues its growth from year to year. In *herbaceous* perennials aerial parts die away in autumn, replaced by new shoots in following year from underground structures, e.g. rhubarb, delphinium; in *woody* perennials, permanent woody stems above ground form starting point for each new year's growth, a characteristic that enables some of them to reach a large size, e.g. shrubs and trees. Cf. *Annual, Biennial, Ephemeral.*

PERFECT. (1) (Of a flower), hermaphrodite; (2) (Of fungi), applied to stage of life-cycle in which spores, formed as result of sexual reproduction are produced.

PERFUSION. Artificial passage of fluid through bloodvessels of an organ or animal. Fluid may be fixative; saline to wash out blood; or may approximate to blood in composition, in order to keep organ alive after isolation from animal.

PERIANTH. Outer part of flower enclosing stamens and carpels, usually consisting of two whorls. In dicotyledons differentiated as an outer green *calyx* consisting of a whorl of *sepals*, and an inner, usually conspicuous, often coloured *corolla* consisting of a whorl of *petals*, e.g. buttercup. In monocotyledons there is usually no differentiation of calyx and corolla, both whorls being alike. In some flowers perianth is reduced, e.g. in stinging nettle where only calyx is present, and grasses where perianth is represented by two tiny scales (*lodicules*); or it is absent altogether, e.g. willow.

PERIBLEM. See *Apical Meristem.*

PERICARDIAL CAVITY (P. SAC). Cavity within which heart lies. In vertebrates, a coelomic space, separated from perivisceral part of coelom (incompletely in elasmobranchs). In Arthropoda and Mollusca, a haemocoelic space supplying blood to heart.

PERICARDIUM. (1) Serous membrane forming wall of pericardial cavity in vertebrates. (2) Sometimes synonymous with pericardial cavity (q.v.).

PERICARP. (Bot.). Wall of an ovary after it has matured into a fruit; may be dry, membranous, or hard, e.g. achene, nut; or fleshy, e.g. berry.

PERICHAETIUM. Distinct whorl of leaves surrounding sex organs in mosses.

PERICLINAL. (Bot.). (Of planes of division of cells), running parallel to surface of plant part. Cf. *Anticlinal.*

PERICYCLE. Tissue of vascular cylinder lying immediately within endodermis, consisting of parenchyma cells and sometimes fibres.

PERIDERM. Cork cambium (phellogen) and its products, i.e. cork and secondary cortex (phelloderm).

PERIGYNOUS. See *Receptacle*.

PERIKARYON. Cell-Body (q.v.).

PERINEUM. Region of body wall between anus and urinogenital openings of placental mammals.

PERIOSTEUM. Layer of connective tissue tightly investing vertebrate bones, to which muscles and tendons are attached. Contains active or (in adult) potential osteoblasts; in embryo, and in adult after fractures, important in formation of bone.

PERIPATUS. See *Onychophora*.

PERIPHERAL NERVOUS SYSTEM. All the rest of the nervous system when the central nervous system is excluded. It consists mainly of *nerves* (q.v.) of various sizes (up to half-inch diameter in man). The major nerves branch repeatedly after leaving the central nervous system to distribute their numerous nerve fibres (perhaps 100,000 in a big vertebrate limb nerve) to the many different effectors and receptors they serve. There are no synapses in the course of most nerves; the nerve fibres are usually continuous right through from the receptors to the central nervous system (in the case of sensory nerve fibres) or from the central nervous system to the effectors (in the case of motor nerve fibres). Though they may run side by side in the nerve, sensory and motor fibres do not usually interact except via the synapses in the central nervous system. But synapses do occur in certain parts of the peripheral nervous system; see *Autonomic Nervous System, Nerve Net*.

PERISPERM. Nutritive tissue surrounding embryo in some seeds, e.g. stitchwort; derived from nucellus (q.v.). Cf. *Endosperm*.

PERISSODACTYLA. Odd-toed ungulates. An order of mammals containing horses, tapirs, and rhinoceros. Walk on their toes which are hoofed, and are characterized by fact that weight-bearing axis of foot lies along third toe. The third toe is usually larger than the others (some of which may disappear); e.g. horses have a very large third toe, and minute second and fourth toes ('splint bones'). One of the two great groups of hoofed mammals (ungulates), other being the Artiodactyla.

PERISTALSIS. Waves of contraction passing along tubular organs, particularly intestine, mixing contents and moving them along. Produced by surrounding coats of smooth muscle.

PERISTOME. (1) (Bot.). Fringe of pointed appendages round mouth of dehiscent capsule in mosses, concerned in spore liberation. (2) (Zool.). Spirally twisted groove leading to mouth of some ciliate Protozoa.

PERITHECIUM. Rounded or flask-shaped fruit-body of certain ascomycete Fungi and lichens, with an internal hymenium of asci and paraphyses and with an apical pore (ostiole) through which the ascospores are discharged.

PERITONEAL CAVITY. Abdominal cavity. Coelomic cavity of mammals posterior to diaphragm, containing liver, spleen, most of gut, and other viscera which almost completely fill it. Term sometimes also used for perivisceral cavity of other vertebrates.

PERITONEUM. Serous membrane lining peritoneal or perivisceral cavity, covering surface of viscera within it, and forming mesentery (q.v.).

PERIVISCERAL CAVITY (COELOM). The main body cavity (q.v.). See *Coelom, Haemocoel.*

PERMANENT TEETH. The second of the two sets of teeth of most mammals. See *Deciduous Teeth.*

PERMEABILITY. Of a membrane, extent to which molecules of a given kind can pass through it. Substances differ greatly in the ease with which they pass through a given membrane, e.g. water or fatty substances pass easily, many ions and proteins with great difficulty, through plasma-membrane (q.v.). Permeability of any biological membrane is itself variable within wide limits. The very numerous membranes of animals and plants, e.g. plasma-membranes or lining of blood capillaries, produce many osmotic effects (see *Osmosis*) and are of great importance in determining local differences in composition of the organism.

PERMIAN. Geological period (q.v.); lasted approximately from 270 till 225 million years ago.

PEROXIDASE. Enzyme (a kind of dehydrogenase, q.v.) which catalyses the oxidation of a substrate by removal of hydrogen which combines with hydrogen peroxide. Occurs particularly in plants.

PES. Foot of hind leg of tetrapod vertebrate. Consists of tarsus, metatarsus and digits. See *Pendactyl Limb, Manus.*

PETAL. One of parts forming corolla of a flower, usually brightly coloured and conspicuous. See *Flower.*

PETIOLE. Leaf-stalk.

PH. A quantitative expression for acidity or alkalinity of a solution, i.e. concentration of hydrogen or hydroxyl ions. Scale ranges from 0 to 14, *p*H7 being neutral, less than 7 acid, more than 7 alkaline. Concentration of hydrogen or hydroxyl ions increases or decreases ten times for each unit of change in *p*H. Theoretically $pH = -\log cH$, where cH is the concentration of hydrogen ions in grams per litre.

PHAEOPHYCEAE. Brown algae (seaweeds). Class of Algae characterized by brown to olive-green colour, chloroplasts containing brown pigment fucoxanthin in addition to chlorophyll. Widely distributed. Thallus filamentous, or complex, differentiated into a disc or root-like basal attachment part, and a stem-like part of varying length, bearing a branched or unbranched, ribbon or leaf-like part, often provided with air bladders, e.g. bladder wrack; up to 60–70 metres in length in largest forms with relatively complex internal structure. Asexual reproduction by fragmentation of thallus or by zoospores. Sexual reproduction isogamous, gametes motile, or anisogamous, by fusion of small, motile, male gamete with large, non-motile egg.

PHAGE. Bacteriophage (q.v.).

PHAGOCYTE. A cell which engulfs into its cytoplasm particles from its surroundings, by a process of flowing all round them called *phagocytosis*. Phagocytes are an important defence mechanism of most Metazoa against invading bacteria. In man and other mammals, polymorphs (q.v.) and macrophages (q.v.) are phagocytic.

PHALANGES. Bones of digit (finger or toe). Each finger has one to five phalanges (more in whales) joined end-to-end in a row, the proximal of each row being jointed to a metacarpal bone. See Fig. 7, p. 175.

PHANEROGAMIA. Name given by early systematic botanists to seed plants (Gymnospermae and Angiospermae); so called because organs of reproduction, e.g. flowers, are clearly evident. Now replaced by Spermatophyta. Cf. *Cryptogamia*.

PHANEROPHYTES. Class of Raunkiaer's Life Forms (q.v.).

PHARYNX. (1) Part of vertebrate gut between mouth and oesophagus into which open glottis in tetrapods, and gill-slits in fish. In man and other mammals, throat and back of nose; partly divided by soft palate into upper (nasal) section and lower (oral or throat) section. Contains sensory receptors setting off swallowing reflex. (2) Part of gut into which gill-slits open internally in ascidians and amphioxus.

PHELLEM. Cork (q.v.).

PHELLODERM. Secondary cortex tissue; formed by cork cambium.

PHELLOGEN. Cork cambium (q.v.).

PHENOLOGY. Study of periodical phenomena of plants, e.g. time of flowering in relation to climate.

PHENOTYPE. The sum of the characteristics manifested by an organism, as contrasted with the set of genes possessed by it (genotype, q.v.). It is possible for organisms to have the same genotype but different phenotypes (owing to environmentally-produced variation); or the same phenotype with different genotypes (e.g. in heterozygotes and homozygotes with a dominant allelomorph; or through incomplete penetrance).

PHLOEM. Vascular tissue that conducts synthesized foods, e.g. sugars, proteins, through the plant. Characterized by presence of sieve-tubes (q.v.) and in some plants, companion cells, fibres, and parenchyma cells. Of two kinds, *primary*, formed by differentiation from procambium and *secondary*, additional phloem produced by activity of cambium.

PHORONIDEA. Small phylum of worm-like animals. Sedentary, live in tubes. Have a ring of ciliated tentacles with which they feed, and superficially resemble certain Polyzoa.

PHOSPHAGEN. Phosphate of creatine, or of arginine (an amino

acid). Reversible splitting off of phosphate from phosphagen plays important part in providing readily available store of energy-rich phosphate for ATP (q.v.) in muscle contraction. Creatine phosphate is the phosphagen of ophiuroids, Acrania, and vertebrates; arginine phosphate of most invertebrates and of tunicates; and both phosphagens occur in echinoids and Hemichordata.

PHOSPHATASES. Enzymes splitting phosphate from its organic compounds (esters).

PHOSPHATIDE. See *Phospholipid.*

PHOSPHOLIPID (PHOSPHOLIPIN, PHOSPHOLIPOID, PHOSPHATIDE). Lipine containing phosphoric acid and nitrogenous base. E.g. lecithin.

PHOSPHORESCENCE. Light given out by living things does not depend on previous illumination and should therefore not be called phosphorescence. It is known as bioluminescence.

PHOSPHORYLATION. Combination (e.g. of sugar) with phosphoric acid.

PHOTONASTY. Response to a general non-directional illumination stimulus, e.g. closing or opening of flowers at night-time, seen respectively in wood sorrel and evening primrose.

PHOTOPERIODISM. Response of plants to relative length of day and night. Growth of plants and formation of flowers and fruits is strongly influenced by length of day, every plant having a particular optimum in this respect. the *photoperiod. Short-day plants* require a daily period of illumination of less than twelve hours for flowering to take place, e.g. cosmos, and if subjected to periods exceeding twelve hours they continue to grow vegetatively, becoming very large, but they will not flower. In contrast to these are *long-day plants,* that need a daily period of illumination of more than twelve hours for flowering to take place, e.g. radish, lettuce, and fail to flower under short-day conditions. The light stimulus is perceived by the leaves and is transmitted, probably in the form of a hormone, to the growing points where it leads to flower initiation. All gradations between the two types exist; some plants, e.g. tomato, flower under widely varying illumination period. Photoperiodic response varies with the age of plant and with temperature.

PHOTORECEPTOR. Receptor (q.v.) detecting light, e.g. vertebrate eye.

PHOTOSYNTHESIS. Synthesis by green plants of organic compounds from water and carbon dioxide using energy absorbed from sunlight. May be summarized by the equation.

$$CO_2 + 2H_2O \xrightarrow[\text{Green plant}]{\text{Light}} (CH_2O) + O_2 + H_2O$$

Initial phase is decomposition (photolysis) of water using light energy absorbed by chlorophyll. Products of photolysis are an oxidant (OH) which eventually yields molecular oxygen and a reductant (H) which is used to reduce carbon dioxide, combined with an acceptor molecule, to the carbohydrate state. Studies with radio active carbon (C^{14}) show that carbon dioxide is combined with a 5-carbon compound to form two phosphoglyceric acid (3 carbon) molecules which are then reduced to the sugar ester triosephosphate with subsequent formation of hexose. From phosphoglyceric acid fatty acids and amino acids may be derived so that fairly direct products of photosynthesis may include not only carbohydrate but fats and proteins. See *Chlorophyll.*

PHOTOTAXIS. Taxis (q.v.) in which stimulus is light.

PHOTOTROPHIC. Of organisms, recently internationally defined as obtaining energy from sunlight. Cf. *Autotrophic*, *Heterotrophic*, *Chemotrophic.*

PHOTOTROPISM (HELIOTROPISM). Tropism (q.v.) in which stimulus is light. E.g. bending of stems of indoor plants towards a window.

PHYCOCYANIN. Photosynthetic blue pigment linked with a protein (globulin) occurring with chlorophyll in blue green and in red Algae.

PHYCOERYTHRIN. Photosynthetic red pigment linked with a protein (globulin) occurring with chlorophyll in red Algae.

PHYCOMYCETES. A group of fungi (q.v.).

PHYLLOCLADE. Cladode (q.v.).

PHYLLODE. Flattened, leaf-like petiole.

PHYLLOTAXY (PHYLLOTAXIS). Arrangement of leaves on the stem, e.g. opposite, whorled, alternate, etc.

PHYLOGENY. Evolutionary history.

PHYLUM. One of the major kinds of group used in classifying animals, e.g. phylum Chordata. Consists of one class or a number of similar classes. Phylum is not often used in plant classification, the term Division being substituted. See *Classification.*

PHYSIOLOGICAL SALINE. Solution of salts in water used for temporarily preserving cells. It must be isotonic (q.v.). The simplest is isotonic sodium chloride (0.9 per cent for mammals), but most cells cannot survive long in a solution of a single salt. A more usual saline is therefore Ringer (q.v.) in which potassium, calcium and magnesium chlorides are also present. For optimum survival, Ringer must be buffered (q.v.) to the right *p*H with phosphates or bicarbonate. Salines do not provide the food requirements of cells (though some contain glucose) so survival in them is limited.

PHYSIOLOGIC SPECIALIZATION (BIOLOGIC SPECIALIZATION). Existence within a particular species of a number of races or forms, which, though indistinguishable in structure, show differences in physiological, biochemical, or pathogenic characters. Of great significance in pathology. E.g. where a number of races of a plant pathogen exist, differing in their pathogenicity towards varieties of the host plant, the problem of breeding a resistant variety is complicated, more so when new races, showing new differences in pathogenicity, may continually be produced. This is the case, e.g. with *Puccinia graminis tritici*, causing Black Stem Rust of wheat, one of the most serious of all plant diseases. More than 200 physiologic races of this fungus have so far been demonstrated.

PHYSIOLOGY. Study of the processes which go on in living organisms.

PHYTOPATHOLOGY. Study of plant diseases.

PHYTOPHAGOUS. Plant-eating.

PHYTOPLANKTON. Plants of plankton (q.v.). Cf. *Zooplankton.*

PIA-ARACHNOID MEMBRANES. Delicate membranes ensheathing vertebrate central nervous system: *pia mater* immediately covering central nervous system, containing many blood-vessels; *arachnoid* outside that, in contact with dura mater (q.v.) and separated from pia by spaces filled with cerebrospinal fluid.

PIGEON'S MILK. Crop milk (q.v.).

PILEUS. Cap of mushroom type of fungus fruit-body, bearing the hymenium on its under surface.

PILIFEROUS LAYER. That part of the epidermis of a root which bears root-hairs. Extends from near the root tip backwards for varying distance depending on species.

PINEAL APPARATUS. Primitively consists of two outgrowths of roof of fore brain lying within skull, one behind the other (though may represent right and left of a pair). Anterior is *parietal organ* which forms an eye-like structure (*pineal eye*, q.v.) in some lizards and *Sphenodon*, a vestigial eye in lamprey (Cyclostomata), and is reduced or absent in other vertebrates. Posterior forms a second, functional eye-like structure in lamprey, and the glandular *pineal body* (q.v.) or *epiphysis* of other vertebrates.

PINEAL BODY (EPIPHYSIS). See *Pineal apparatus.* Its function is unknown, but often assumed to secrete a hormone.

PINEAL EYE. See *Pineal apparatus.* Pineal eye is a vesicle, formed as diverticulum of brain. In most complete form outer layer of vesicle becomes a lens (unlike vertebrate paired eyes, see *Lens*), inner layer a retina connected by nerve to brain. There is a gap in skull above the eye, and skin covering is almost transparent. Nerve however usually degenerates in adult reptile, though not in lamprey. Function unknown.

PINNA. (1) Leaflet, primary division of a leaf. (2) See *Ear, outer.*

PINNIPEDIA. Seals, walruses, etc. Specialized aquatic branch of the order Carnivora of placental mammals.

PISCES. Fish. Formerly regarded as a single class of vertebrates, are now resolved into (usually) four distinct classes; Placodermi, Chondrichthyes, Actinopterygii, Choanichthyes. The latter two are united by some zoologists as the Osteichthyes.

PISTIL. (Bot.). (1) Each separate carpel of an apocarpous gynoceium.

(2) The gynoecium as a whole whether it is apocarpous or syncarpous.

PISTILLATE. (Of flowers), possessing carpels but not stamens; i.e. female. Cf. *Staminate.*

PITH. See *Medulla.*

PITHECANTHROPUS. Java ape-man. Lower Pleistocene. Brain of human type but small; walked erect; ape-like in heavy brow ridges and absence of chin prominence. See *Sinanthropus.*

PITHECOIDEA. Anthropoidea (q.v.).

PITRESSIN. Extract of posterior lobe of pituitary, causing when injected constriction of capillaries and arterioles (which raises arterial pressure); and diminished urine formation (see *Anti-diuretic hormone*).

PITS. Small, sharply defined areas in walls of plant cells that remain thin while the rest of the wall becomes thickened to a greater or lesser extent; coinciding in position with pits in walls of adjacent cells, and facilitating communication between them. In mature cells pits appear as depressions in the wall closed by original unthickened portion of cell wall (pit membrane). There are two kinds of pit, *simple* pits as described, characteristic of living cells, stone cells, and some fibres; and *bordered* pits, characteristic of xylem vessels and tracheids, in which the pit cavity is partly enclosed by over-arching of the cell wall and the pit membrane may possess a central, thickened impermeable area (*torus*), which closes aperture of pit if pit membrane is laterally displaced.

PITUITARY BODY P. GLAND, HYPOPHYSIS CEREBRI). Endocrine gland beneath floor of brain, within skull, of vertebrates. Many hormones, all proteins, have been extracted from pituitary, but it is uncertain how many are really secreted as separate hormones, or whether some have been artificially separated during extraction. Two main regions, *anterior lobe* (or *anterior pituitary*) and *posterior lobe* (or *posterior pituitary*), separated by a cleft. From anterior lobe are obtained hormones which control activities of other endocrine glands, e.g. thyrotrophic, gonadotrophic and corticotrophic hormones controlling hormone production of thyroid, gonads, and adrenal cortex respectively; but direct-acting ones as well, e.g. prolactin, and a growth-controlling hormone. From posterior lobe are obtained anti-diuretic hormone and oxytocin. Pituitary itself is closely connected with immediately overlying part of brain (hypothalamus) and posterior lobe at least is strongly influenced by it. Embryologically has a double origin, from hypophysis in strict sense (q.v.) and infundibulum (q.v.); anterior lobe from former, posterior lobe from latter.

PITUITRIN. Extract of posterior lobe of pituitary having activity of both oxytocin and pitressin.

PLACENTA. (Bot.). Part of ovary wall (fused margins of carpel(s)) on which ovules are borne. (Zool.). Organ, consisting of embryonic and maternal tissues in close union, by which embryo of viviparous

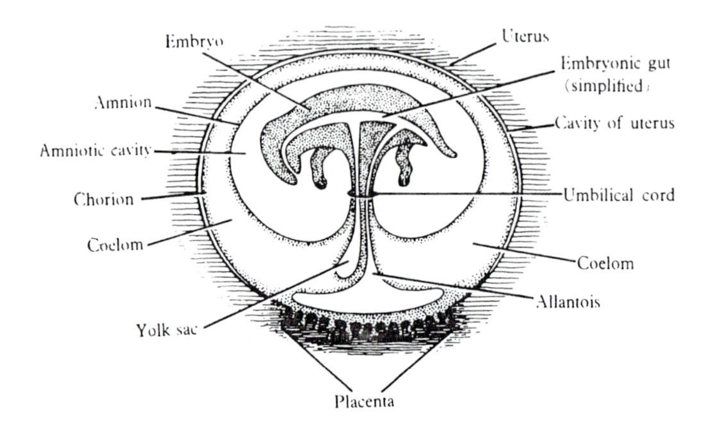

Fig. 8. Diagrammatic section through foetus of a mammal in the uterus, showing relation to placenta, and extra-embryonic membranes. Mesoderm stippled. Arrangement of extra-embryonic membranes is fundamentally the same in all amniotes; though relative size of allantois and yolk-sac vary, and in birds and reptiles placenta is absent, and chorion is enclosed by albumen and shell instead of by uterus.

animal is nourished. In placental mammals the embryonic tissue concerned is the chorion, with blood vessels supplied via allantois or occasionally via yolk-sac. In ungulates the trophoblast of the chorion merely fits closely against endometrium of uterus; but in most mammals the endometrium becomes much altered (decidua, q.v.) and in parts eroded, so that the trophoblast is in contact with maternal blood-vessels (carnivores) or maternal blood (other mammals including man). There is exchange of small molecules through the trophoblast between maternal and embryonic blood in the mammalian placenta; oxygen and food-substances enter the embryo, urea and carbon dioxide leave it. Only minute amounts of proteins are exchanged. The placenta produces hormones (gonadotrophic, oestrogen, progesterone) which keep uterus adapted to the pregnancy. At birth the umbilical cord (q.v.), which connects placenta to embryo, is broken, and the placenta is discharged from the uterus after the embryo. See Fig. 8.

PLACENTALIA (EUTHERIA, PLACENTAL MAMMALS). A subclass of Mammalia, containing the great majority of living mammals. Characteristic of placentals is that the embryo develops in the maternal uterus, attached to the maternal tissues by a highly organized placenta. The cerebral cortex is larger and more complex than in the other mammals.

PLACENTATION. (Bot.). Type of arrangement of placentas in a syncarpous ovary; (*a*) parietal; carpels are fused only by their margins, placentas then appearing as internal ridges on ovary wall, e.g. violet; (*b*) axile; margins of carpels fold inwards, fusing together in centre of ovary to form a single, central placenta. Ovary is divided into as many compartments (loculi) as there are carpels, e.g. lily; (*c*) free central; placenta arising as a central upgrowth from ovary base, e.g. primrose.

(Zool.). Type of arrangement of placenta in mammals.

PLACODERMI (APHETOHYOIDEA). Fish-like class of aquatic vertebrates, entirely fossil and mainly Devonian (350–400 million years ago). Differ from other fish classes in having a functional first gill slit just behind the mouth on each side (see *Spiracle*) and autostylic jaw suspension (q.v.).

PLACOID SCALE. Denticle (q.v.).

PLAGIOCLIMAX. Type of plant community the composition of which is more or less stable, in equilibrium under existing environmental conditions, but which has not achieved the natural climax under those conditions due to action of biotic factors, e.g. grassland under continuous pasture. Cf. *Climax.*

PLAGIOGEOTROPISM. Orientation of plant part by growth curvature in response to stimulus of gravity so that its axis makes an angle other than a right-angle with the line of gravitational force; exhibited, e.g. by branches of a main root which make an acute angle with the vertical. Also used in a wider sense to mean orientation of an organ so that its axis makes a constant angle with the vertical and including diageotropism (q.v.) as a special type.

PLAGIOSERE. Succession (q.v.) of plants deflected into a new course by biotic factor or factors. Cf. *Prisere.*

PLANKTON. Animals and plants of sea or lake which float or drift almost passively. They are mostly very small; the smallest are *nannoplankton*, e.g. diatoms. Plankton occurs mainly near the surface, where the plants get suitable illumination. Of great ecological and economic importance, providing food for fish and whales.

PLANTIGRADE. Walking on ventral surface of whole foot, i.e. of metacarpus or metatarsus and of digits (cf. *Digitigrade*, *Unguligrade*). E.g. man is plantigrade.

PLASMA. Blood plasma (q.v.).

PLASMAGEL. See *Ectoplasm.*

PLASMAGENE. Particle in cytoplasm having the properties of a gene in that it is self-reproducing, shows inheritance from cell to daughter-cell, and affects the character of the cell bearing it; but unlike a gene in that it is not inherited in mendelian fashion through the gametes.

PLASMA-MEMBRANE (CELL-MEMBRANE). Extremely thin membrane (perhaps about one-hundredth micron thick) mainly of fat and protein covering surface of all cells (or protoplast in plants). Responsible for restricted penetration of many substances into interior of cell (see *Permeability*). Severe damage to membrane at once destroys cell. In plants plasma-membranes occur both in contact with cell-wall (external, ectoplast) and bordering vacuole (internal, tonoplast).

PLASMA PROTEINS. Dissolved proteins of vertebrate blood plasma (about seven per cent of weight of human blood) which are responsible for holding fluid in blood-vessels by osmosis (see *Capillary*). To them is due the value of transfusion of plasma in increasing volume of circulating blood. Amongst plasma proteins are antibodies and blood-clotting (q.v.) substances.

PLASMASOL. See *Endoplasm.*

PLASMODESMATA. Extremely fine cytoplasmic threads passing through cellulose walls of living plant cells and forming a link between cytoplasm of adjacent cells. See *Cell-wall.*

PLASMODIUM. (1) Vegetative stage of Myxomycetes (q.v.). Multinucleate, amoeboid mass of protoplasm bounded by plasma membrane, without a definite size or shape. Cf. *Coenocyte, Syncytium.* (2) Generic name of malaria parasite, a Sporozoan.

PLASMOLYSIS. (Bot.). Shrinkage of cell protoplasm away from its cellulose wall when placed in hypertonic (q.v.) solutions, due to osmotic withdrawal of water from its large central vacuole.

PLASTIDS. Small, variously shaped, self-propagating protoplasmic inclusions occurring in cytoplasm of plant cells, classified according to colour as *leucoplasts* and *chromoplasts.* Leucoplasts are colourless plastids in which starch is deposited. They occur in cells that are not normally exposed to light, e.g. in deep-seated tissue, and in tissue of underground organs, and may develop chlorophyll on exposure to light. Chromoplasts are coloured plastids; of two kinds, (1) containing chlorophyll, *chloroplasts*, to which green plants owe their colour and without which photosynthesis cannot take place; (2) red, orange, or yellow plastids containing carotin and/or xanthophyll, found in many flowers and fruits, e.g. tomato. The term chromoplast is often confined to this second group of pigmented plastids.

PLASTOGENE. Plastid (q.v.) carrying a factor having, like a plasmagene (q.v.), properties of a gene (q.v.) but, unlike a gene, not inherited in mendelian fashion.

PLATELET. Blood platelet (q.v.).

PLATYHELMINTHES (FLATWORMS). Phylum containing eddy-worms, flukes, tapeworms. Resemble coelenterates in that the gut, when present, has only one opening, and there is no coelom, nor blood system; but differ in that they are bilaterally, not radially, symmetrical; have bulky parenchymatous tissue derived from the mesoderm; flame cells; and a very complex hermaphrodite reproductive system. Includes classes Turbellaria, Trematoda, Cestoda; some parasites of great economic and medical importance.

PLATYRRHINES. The New World monkeys, members of group Platyrrhini (order Primates). Characterized by broad nasal septum. Often have prehensile tails. E.g. marmoset, howler. See *Catarrhine*.

PLECOPTERA. Stone-flies. Small order of exopterygote insects with aquatic nymphs.

PLECTENCHYMA. Tissue of fungal hyphae. Of two types, *prosenchyma*, in which component hyphae, loosely woven, can be recognized as such and *pseudoparenchyma*, consisting of closely packed cells, no longer distinguishable as hyphae, resembling parenchyma (q.v.) of higher plants.

PLEISTOCENE. Geological period (q.v.); lasted approximately from one million till ten thousand years ago. During it four major ice ages occurred. Succeeded by Recent period.

PLEROCEROID. Larva of certain tapeworms in which there is no bladder, the body being solid (cf. *Cysticerus*, *Cysticercoid*). See *Onchosphere*.

PLEROME. See *Apical meristem*.

PLESIOSAURIA. A fossil order of Reptilia; lived during Jurassic and Cretaceous. Marine; swam by large paddle-like limbs; long neck, short tail; up to 50 ft in total length.

PLEURA. Serous membrane lining pleural sac, and covering surface of lung, in mammals.

PLEURAL CAVITY (P. SAC). Coelomic space surrounding lung in mammal, separated from rest of perivisceral coelom by diaphragm. There is a pair of pleural sacs separated from each other by mediastinum and pericardial sac. Actual space of sac is normally a thin layer of fluid between pleura of lungs and body-wall, which are almost in contact.

PLIOCENE. Geological period (q.v.), sub-division of Tertiary, lasted approximately from ten till one million years ago.

PLUMULE. (Zool.). Down feather. Temporary feather of nestling bird, persisting in some adult birds between contour feathers. No barbules (q.v.). (Bot.). Terminal bud of embryo in seed plants.

PLUTEUS. Larva of echinoid or ophiuroid; ciliated, planktonic.

PNEUMATOPHORE. Special root branch produced in large numbers by some vascular plants growing in water or in tidal swamps, e.g. mangrove; grows erect, projecting into the air above and contains well developed intercellular system of air spaces in communication with atmosphere through pores on aerial portion.

PNEUMOGASTRIC NERVE. Vagus nerve (q.v.).

PODIA. Tube-feet (q.v.).

POIKILOTHERMIC. 'Cold-blooded', but actually with varying body temperature, which approximately follows that of surroundings. Cf. *Homoiothermic*. Characteristic of all animals except birds and mammals. Aquatic poikilotherms keep very close to temperature of surrounding water. Terrestrial ones may differ considerably from surrounding air, either cooling by evaporation, or heating by sun's radiation or muscular exercise.

POLAR BODY (POLOCYTE.) Minute cell produced during development of oocyte (q.v.) containing one of the nuclei derived from first or second division of meiosis (q.v.) but practically no cytoplasm.

POLLEN. Microspores of seed plants (Gymnospermae and Angiospermae) each containing a greatly reduced male gametophyte. Carried by wind or insects, or more rarely by water, to ovules (Gymnospermae) or to the stigmas of the carpels (Angiospermae), where they germinate with production of pollen tubes carrying male nuclei to eggs within ovules.

POLLEN ANALYSIS (PALYNOLOGY). Method of reconstructing floras of the past based on observations on pollen grains (and other spores) preserved in ancient peat and sedimentary deposits. The resistance to decay of the outer coats of pollen grains with their distinctive sculpturing makes possible both qualitative and quantitative estimates of species occurring thousands of years ago.

POLLEN TUBE. Tube formed on germination of pollen grain that carries male gametes to egg. In flowering plants each pollen tube penetrates tissues of stigma and style into ovary. After entering ovary pollen tube grows towards an ovule, passing down micropyle to nucellus and finally penetrating nucellus tissues into embryo sac where it ruptures at the tip, setting free two male gametes, one of which fuses with egg nucleus. The second male nucleus fuses with the central fusion nucleus; the resulting nucleus is known as the endosperm nucleus and from it arises endosperm.

POLLEX. 'Thumb' of pentadactyl fore-limb. The lateral digit on side of fore-foot corresponding to the radius. Often shorter than other digits.

POLLINATION. Transference of pollen from anther to stigma. See *Anemophily*, *Entomophily*.

POLLINIUM. Coherent mass of pollen grains, as in orchids.

POLOCYTE. Polar body (q.v.).

POLYADELPHOUS. (Of stamens) united by their stalks (filaments) into several groups, e.g. St John's Wort. Cf. *Monadelphous*, *Diadelphous*.

POLYCHAETA. Order of annelids (class Chaetopoda) including bristleworms, tube-worms, fan-worms. Marine; chaetae numerous, borne on projections of body (parapodia); usually well marked head with special appendages; unisexual; fertilization external. Cf. *Oligochaeta*.

POLYEMBRYONY. (Bot.). Formation of more than one embryo per ovule, often by vegetative budding from pro-embryo. (Zool.). Formation of more than one embryo per zygote by fission at some early stage of development. E.g. an armadillo (*Dasypodus*) always has identical quads from a single ovum. Most striking cases are in parasitic Hymenoptera, where up to 2,000 embryos may spring from one zygote. Monozygotic twins (q.v.) are the simplest form of polyembryony.

POLYGENES. Genes with individually a very small effect on phenotype differences, systems of which are associated in producing quantitative variation of particular characters, e.g. stature. Quantitative variation is also environmentally produced. See *Blending Inheritance*.

POLYMORPH (POLYMORPHONUCLEAR LEUCOCYTE, GRANULOCYTE). Kind of white blood cell of vertebrates. 7–9 microns diameter in man, with darkly staining nucleus constricted into a number of lobes (lobation of nucleus may be absent in some vertebrates). Cytoplasm usually contains conspicuous granulations. Polymorphs form 65–75 per cent of white blood cells in man. Actively phagocytic especially when, in inflammation (q.v.), they invade the tissues. Produced continually in bone marrow (myeloid tissue, q.v.) each polymorph having only a short life. In man, granulations of most polymorphs (neutrophils) stain preferentially with neither acid nor basic dyes, those of some (eosinophils, q.v.) stain strongly with acid dyes, those of a few (basophils) with basic dyes.

POLYMORPHISM. Having several different forms. Particularly used of distinct kinds of individual belonging to one species, occurring in fairly constant proportions within a freely interbreeding population. E.g. within some species of butterfly there co-exist individuals which mimic other species and individuals which do not; another example is human blood groups. Dimorphism of the two sexes is exactly analogous. The different types are commonly controlled by different genes of one allelomorphic set. Insect castes (q.v.) are a special case of polymorphism. (Of fungi) having several different spore forms, e.g. rust fungi.

POLYP (HYDRANTH). Sedentary form of coelenterate, e.g. *Hydra*, sea-anemone. Cylindrical trunk, fixed at one end; mouth at other end, surrounded by ring of tentacles. Many polyps reproduce themselves by budding, and may form colonies. Some reproduce sexually, fertilized eggs giving rise to new polyps (*Hydra*, sea-anemone). Others do not produce sexually, but produce medusae (q.v.) by budding and these reproduce sexually, their fertilized eggs giving rise to new polyps (e.g. hydroids; jelly-fish).

POLYPEPTIDE. Peptide (q.v.) formed of three or more amino-acids, but not as many as make a peptone.

POLYPETALOUS. (Of a flower), with petals free from one another, e.g. buttercup. Cf. *Gamopetalous*.

POLYPHYLETIC. A group of species classified together is polyphyletic when some of its members have had quite distinct evolutionary histories, not being descended from a common ancestor which was also a member of the group; so that, if the classification is to correspond with phylogeny, the group should be broken up into several distinct groups. E.g. phylum Polyzoa is generally regarded as being polyphyletic and is therefore better split into two phyla Endocprocta and Ectoprocta. Cf. *Monophyletic*.

POLYPIDE. Member of a colony of Polyzoa.

POLYPLOID. Having three or more times the haploid number of chromosomes. Polyploid individuals, sub-species and species are fairly common in plants (especially angiosperms) but rare in animals. This is mainly because polyploid organisms are sterile when crossed with diploids, and a polyploid suddenly arising in a diploid population can therefore only reproduce vegetatively, parthenogenetically, or by self-fertilization. These ways are commonly open to plants, but not commonly to animals. See *Allotetraploid.* Polyploid cells and tissues occur in otherwise diploid organisms (see *Endomitosis*).

POLYPTERUS. Primitive actinopterygian fish of tropical Africa, with well developed lungs and ganoid scales.

POLYSACCHARIDE. Carbohydrate produced by combination of many molecules of monosaccharide; e.g. starch, cellulose. Some kinds, e.g. chitin, have other constituents besides such simple sugars. Large, often fibrous, molecules. Important structural and reserve energy-rich material in organisms.

POLYSEPALOUS. (Of a flower), with sepals free one from another, e.g. buttercup. Cf. *Gamosepalous.*

POLYSPERMY. Penetration of numerous sperm into one ovum at time of fertilization. Occurs normally in very yolky eggs (e.g. shark, bird); only one sperm nucleus fuses with egg nucleus, the rest taking little further part in development. May occur abnormally in other eggs, upsetting development.

POLYZOA (BRYOZOA). Sea-mats, corallines. Phylum of small, aquatic, usually fixed and colonial animals, superficially resembling hydroid coelenterates but considerably more complex. Have ciliated tentacles with which they feed; anus; coelom; some have horny or calcareous skeletons. Contains two classes, Ectoprocta and Endoprocta, which are sometimes regarded as distinct phyla.

POME. False 'fruit' greater part of which is developed from receptacle of flower, not from ovary; e.g. apple, edible, fleshy part representing receptacle and the core, ovary.

POPULATION CYCLE. Rise and fall of population numbers with regular periodicity. E.g. the 10-year cycle of the snowshoe rabbit and other mammals and birds of N. America.

PORIFERA. Sponges. Non-mobile animals consisting of many cells but built on a different plan from the Metazoa; hence grouped separately from all other multicellular animals as Parazoa. They have for instance, no nervous system, and possess collar cells (choanocytes, q.v.) which are found otherwise only in a certain group of flagellates, the choanoflagellates, from which sponges have probably evolved. Currents of water are drawn into the body through small pores (ostia) and passed out through larger pores (oscula), food particles being collected en route. In most sponges the body is supported by skeleton of lime, silica, or spongin (a protein).

PORTAL VEIN. Vein carrying blood from one capillary network to another. See *Hepatic portal system*, *Renal portal system*.

POSITION EFFECT. Influence, on characters produced by a given gene, of change of its position in the chromosomes relative to other genes (e.g. by translocation).

POSTCAVAL VEIN. Vena Cava Inferior (q.v.).

POSTERIOR. (Bot.). Of lateral flowers, part nearest the main axis. See *Floral diagram*. (Zool.). Situated at or relatively nearer to the hind end, i.e. usually the end directed backwards when the animal is in motion. (In human anatomy, the posterior side is the back, which is equivalent to *dorsal* side of other mammals.)

POSTERIOR ROOT. Nerve-root, synonymous with dorsal root (q.v.).

PRECAVAL VEIN. Vena Cava Superior (q.v.).

PREFORMATION. Hypothetical pre-existence of entire adult diversity of structure in the fertilized egg; embryonic development supposedly consisting merely of enlargement and manifestation of this structure. Cf. *Epigenesis*.

PREMAXILLA. Dermal bone forming front part of upper jaw in most vertebrates, bearing teeth (incisors in mammals). Forms most of upper beak in birds.

PREMOLARS. Those crushing cheek teeth of mammals which (unlike molars) have predecessors in deciduous (milk) set of teeth. They have usually more than one root, and a pattern of ridges and projections on biting surface. See *Dental Formula*.

PRESUMPTIVE. (Zool.). Embryological term applied to tissues before their differentiation, meaning 'becoming in the course of normal development'. E.g. if a tissue is 'presumptive epidermis' it means that in normal development it would become epidermis, regardless of whether or not it is already determined (q.v.) as epidermis. A *presumptive region* is part of an embryo consisting of all the cells which will, in normal development, become a particular organ or tissue. Maps of presumptive regions have been prepared for early stages of many species.

PRIMARY MERISTEM. Region of active cell division that has persisted from origin in embryo or young plant, e.g. growing points (apical meristems) of root and stem, cambium. Cf. *Secondary Meristem*.

PRIMATE. A member of the Primates, an order of placental mammals, containing man, apes, and monkeys. A primarily arboreal order, rather primitive in structure. Usually climb by grasping; nails commonly present instead of claws, and big toe and often thumb usually well-developed and opposable to other digits. Tree-life has also involved particularly well-developed eye-sight, often with binocular vision (see *Fovea*). Relatively large brain. First appear at beginning of Tertiary. Includes three sub-orders, Lemuroidea, Tarsioidea, and Anthropoidea.

PRIMITIVE. At an early stage in the evolutionary history of a given group; or similar to an organism (or part thereof) at such an early stage.

PRIMITIVE STREAK. Longitudinal thickening in disc-like early embryo of bird or mammal during the stage of gastrulation (q.v.). The streak is produced by the accumulation of mesoderm as this material moves from its originally superficial position into the interior of the embryo. See *Blastopore.*

PRIMORDIAL MERISTEM. Promeristem (q.v.).

PRISERE. Primary sere. Complete, natural succession of plants, from bare habitat to climax. Cf. *Plagiosere.*

PROBOSCIDEA. Elephants. An order of placental mammals, modern ones characterized by proboscis (trunk), large size, pillar-like legs, greatly lengthened incisors (tusks), and huge grinding molars only two pairs of which are in use at a time.

PROCAMBIUM. Tissue of elongated, narrow cells grouped into strands, differentiated in the plerome (q.v.) just behind growing point in stem and root of vascular plants; gives rise by further development to vascular tissue.

PROCHORDATA. Protochordata (q.v.).

PROCTODAEUM. An intucking of ectoderm of embryo meeting endoderm of posterior end of gut, forming anus or cloacal opening.

PRO-EMBRYO. In seed plants, group of cells formed by initial divisions of fertilized egg-cell which by further development is differentiated into suspensor and embryo proper.

PROGESTERONE. Hormone (a steroid) secreted by corpus luteum (q.v.) of mammalian ovary, responsible for preparing reproductive organs for pregnancy, as in luteal phase of oestrous cycle (q.v.); and for maintaining uterus in special state for nourishment and protection of embryo during pregnancy, when it is also produced by placenta.

PROGESTOGEN. General term for substance with progesterone-like effects.

PROGLOTTIS (PL. PROGLOTTIDES). One of the string of segments of which a tapeworm consists; when mature each has at least one set of reproductive organs.

PROLACTIN. See *Lactogenic hormone.*

PROLEGS. Stumpy appendages with no joints, on abdomen of caterpillars.

PROLIFERATION. Growth by active cell-division.

PROMERISTEM (PRIMORDIAL MERISTEM). Apex of growing point (apical meristem) of vascular plants, consisting of actively dividing, uniformly shaped cells. See *Apical meristem.*

PRONATION. Position, or rotation towards the position, of fore-limb such that fore-foot (hand) is twisted through 90 degrees relative to elbow, the radius and ulna being crossed. Men and other primates can untwist the fore-arm (*supination*).

PRONEPHROS. The first part of kidney of vertebrates to arise in embryonic development. It appears at border of somites and lateral plate, just behind heart, as a number (varying according to species) of nephrons (q.v.) segmentally arranged (i.e. one per somite). Commonly the nephrons communicate with the coelom, and they join a tube (*pronephric duct*) which grows back from their neighbourhood to the cloaca, putting coelom into communication with exterior (see *Coelomoduct*). Functional kidney of larvae of anamniotes (e.g. tadpole); but small and non-functional in those without larvae (e.g. elasmobranchs) and vestigial in amniotes. Its duct becomes Wolffian duct (q.v.). See *Mesonephros*, *Metanephros*.

PRONUCLEUS. Sperm nucleus (male pronucleus) after its entry into ovum at fertilization, but before fusion with ovum nucleus; or ovum nucleus (female pronucleus) after completion of meiosis but before fusion with sperm nucleus. Pronuclei are haploid.

PROPHASE. Initial stage of mitosis (q.v.) or meiosis (q.v.) during which chromosomes appear within the nucleus, and in meiosis undergo pairing. See Figs. 6A, 6B, pp. 147, 148.

PROPRIOCEPTOR. (1) Receptor (q.v.) which detects position and movement, e.g. kinaesthetic receptors such as those of muscle of vertebrates and skeleton of insects recording passive stretch, or contraction; or balancing organs of internal ear. Usually do not show sensory adaptation. Cf. *Interoceptor*. (2) In wider sense, receptor which detects changes within the body other than those caused by substances taken into gut or respiratory tract; including, e.g. deep pain receptors, and receptors detecting pressure in blood-vessels, as well as those included under (1).

PROSENCEPHALON. Fore-brain (q.v.).

PROSENCHYMA. See *Plectenchyma*.

PROSTATE GLAND. Gland of male reproductive system of mammals which contributes substances to semen, of unknown function. Its size and secretory powers are controlled by androgens.

PROSTHETIC GROUP. Non-protein substance when it is combined with a protein, e.g. nucleic acid in nucleo-protein.

PROTANDROUS. (Of flowers), anthers maturing before carpels, e.g. dandelion. See *Dichogamy*. (Of animals), producing first sperm, and then eggs, e.g. certain roundworms. Cf. *Protogynous*.

PROTEASE. See *Proteolytic enzyme*.

PROTEIN. Very complex organic compound, composed of numerous amino-acids (q.v.). Proteins of innumerably many different kinds are present in all living things, making a considerable proportion of their dry weight. A protein molecule is made of hundreds or thousands of amino-acid molecules joined together by peptide links (q.v.) into one or more chains, which are variously folded. There are 20 different kinds of amino-acid commonly found in proteins, and most of these usually occur in any one protein molecule; they are arranged in the chain in a sequence which seems to be exactly the same in all molecules of a given kind of protein. The possible different arrangements of the amino-acids are evidently practically infinite, and the diversity is fully exploited by living things, every species having kinds of protein molecule peculiar to itself. A protein molecule is very large (most have a molecular weight from about 20,000 up to several million), and dissolved proteins form therefore colloidal solutions. Proteins are not soluble in fat solvents. Many are soluble in water or dilute salt solutions (e.g. globulins); others, with elongated (*fibrous*) molecules, are insoluble in these solvents (e.g. scleroproteins, myosin). Proteins are synthesized from amino-acids by all living things; the precise sequence of the amino-acids seems to be determined by the sequence of nucleotides in nucleic acids; only autotrophic organisms make them from inorganic substances. They are destroyed by proteolytic enzymes (q.v.) (see *Peptone, Polypeptide*). Proteins are frequently combined with other substances, especially nucleic acid (nucleo-proteins), carbohydrates (muco- or glycoproteins), fats (lipoproteins).

PROTEINASE (ENDOPEPTIDASE). See *Peptidase.*

PROTEOLYTIC ENZYME. Any enzyme taking part in breaking down of proteins. A system of several peptidases (q.v.) is necessary to break proteins down to constituent amino acids; and such frequently occurs, e.g. in plant and animal cells (see *Cathepsin*), and in plant seeds and animal digestive juices.

PROTHALLUS. Independent, gametophyte plant of Pteridophyta (ferns and related plants). Small, green parenchymatous thallus bearing antheridia and archegonia; shows little differentiation. Usually prostrate on soil surface attached by rhizoids; may be subterranean, mycotrophic.

PROTHORAX. Most anterior of the three segments which make up insect thorax. Bears a pair of walking-legs, but no wings.

PROTHROMBIN. See *Blood clotting.*

PROTISTA. Term sometimes used for all unicellular organisms; plants (i.e. unicellular Algae), animals (i.e. Protozoa), and Bacteria.

PROTOCHORDATA (PROCHORDATA). Term sometimes applied to a heterogeneous group of animals of phylum Chordata, related to the vertebrates, which they resemble in possessing gill slits, notochord, and dorsal hollow nerve cord, or at least traces of these. Includes Acrania, a group consisting of various species of amphioxus, a small animal which looks fish-like, but has no skeleton, brain, or eyes; Urochordata (sea-squirts, etc.), which in the adult form have little resemblance to vertebrates: and Hemichordata (e.g. *Balanoglossus*) worm-like forms whose vertebrate affinities are much more remote.

PROTOGYNOUS. (Of flowers), carpels maturing before anthers, e.g. plantain. See *Dichogamy.* (Of animals), producing first eggs, and then sperms. Cf. *Protandrous.*

PROTONEMA. Branched, multicellular, filamentous or less commonly thalloid structure produced on germination of spore in members of Bryophyta from which new plants develop as buds.

PROTONEPHRIDIUM. Organ thought to be excretory, present in Platyhelminthes, Nemertea, Rotifera, and some trochophore larvae; consisting of one or more flame cells (q.v.), at the inner end of tubes which open to the exterior.

PROTOPLASM. Substance within and including plasma-membrane of a cell (q.v.) or protoplast (q.v.) but usually taken to exclude large vacuoles, or masses of secretion, or ingested material. In animals and plants differentiated into nucleus (q.v.) and cytoplasm (q.v.).

PROTOPLAST. (Bot.). The actively metabolizing part (i.e. the protoplasm) of a cell (q.v.) as distinct from the cell-wall (q.v.). Equivalent to what a zoologist means by a cell.

PROTOPTERUS. African lungfish (Dipnoan).

PROTOSTELE. Most simple, primitive type of stele (q.v.), consisting of solid, central core of xylem surrounded by cylinder of phloem. Present in stems of certain ferns, club mosses; almost universally in roots.

PROTOTROPH. Strain of a micro-organism (alga, bacterium or fungus) having no nutritional requirements in addition to those of the wild type (q.v.) from which it was derived. Cf. *Auxotroph.*

PROTOXYLEM. First xylem elements to be differentiated from procambium (q.v.); extensible. Described as *endarch* when internal to the later-formed metaxylem, e.g. in stems of angiosperms; *exarch* when external to metaxylem, e.g. in roots; and *mesarch* when surrounded by metaxylem, e.g. in stems of ferns.

PROTOZOA. Group (phylum or subkingdom) of animals differing from the rest (Metazoa and Parazoa) in consisting of one cell only, but resembling them and plants, and differing from Bacteria, in having at least one well-defined nucleus. Contains classes Flagellata, Rhizopoda, Ciliophora, Sporozoa. Unicellularity does not necessarily imply close relationship; the Fungi and Algae also include unicellular forms. The status of the Flagellata (q.v.) is particularly complicated. Protozoa are ubiquitous, and of great significance in the economy of nature; some parasitic forms are important economically and medically (e.g. malaria parasite).

PROVENTRICULUS. (1) Anterior part of stomach of birds, where digestive enzymes are secreted; the posterior part of stomach being the gizzard. (2) Of some invertebrates (Crustacea, Insecta) the gizzard.

PROXIMAL. Situated towards point of attachment. Cf. *Distal.*

PSEUDOGAMY. Development of ovum into a new individual as a result of stimulation by a male gamete, whose nucleus however does not fuse with that of ovum, and contributes nothing to hereditary constitution of embryo. Occurs in some nematodes, and in some higher plants.

PSEUDOPARENCHYMA. See *Plectenchyma*.

PSEUDOPODIUM. Temporary protrusion of the cell, associated with flowing movements of protoplasm, occurring in rhizopod and other Protozoa, white blood cells, etc. Serves locomotion and feeding (phagocytosis). Physiology of such movement is not understood.

PSEUDOPREGNANCY. Occurrence in female mammal of changes closely resembling those of pregnancy, but in absence of embryos; due to hormone secretion by corpus luteum (q.v.) of ovary. Occurs in species where active corpus luteum is formed as a result of copulation, when such copulation is sterile (e.g. rabbit, mouse); or where normal oestrus cycle (q.v.) includes a pronounced luteal phase (e.g. bitch).

PSILOPHYTALES. Order of extinct, palaeozoic Pteridophyta (q.v.) that flourished in the Devonian (q.v.). Oldest land plants known, with very primitive organization. Small, rhizomatous plants with dichotomously branched aerial shoots a few mm. to about a cm. in diameter, leafless or with many small leaves, with simple vascular tissue; bearing apical sporangia.

PSILOTALES. Order of Pteridophyta (q.v.) related to extinct and most primitive Pteridophytes (Psilophytales, q.v.); represented by two genera, *Psilotum* and *Tmesipteris*. Usually found as subtropical or tropical epiphytes. Small, rhizomatous plants with dichotomously branched aerial shoots. Sporangia borne in groups of two or three in axils of small, two-lobed sporophylls. Homosporous, gametophytes subterranean, lacking chlorophyll, mycotrophic; bearing antheridia and archegonia scattered indiscriminately.

PSOCOPTERA. Booklice. Order of exopterygote insects. Small, some wingless. Common booklouse, found in damp rooms, feeds mainly on fungus.

PTERIDOPHYTA. Division of plant kingdom comprising ferns, horsetails, clubmosses, etc. A prominent group of the Carboniferous era, when tree-like forms were common; living representatives forming a small, widely distributed group attaining its greatest development in the tropics. Largely terrestrial; characterized by possession of true stems, leaves, and roots; a well-developed vascular system (but no cambium); sporangia borne on leaves (*sporophylls*) and by separate, small, inconspicuous sexual plants (*prothalli*), bearing antheridia and archegonia. Sporophylls resembling ordinary foliage leaves, e.g. bracken, or much modified, arranged in a cone, e.g. horsetail. The sporangia liberate spores which germinate to give rise to prothalli. Fertilization of egg in archegonium is effected by male gametes motile by flagella. Zygote develops into a new plant. Alternation of generations (q.v.) is well marked. The fern plant, the sporophyte, is the dominant generation; the prothallus is the gametophyte. Contrasted with lower plants the outstanding feature of the pteridophytes is the development of an independent sporophyte, with roots, stems, leaves, and well-developed vascular system. In most ferns the sporophyte gives rise to only one kind of spore and one kind of gametophyte (prothallus) bearing both antheridia and archegonia; they are *homosporous*. In other, *heterosporous* members, e.g. *Selaginella*, the sporophyte produces small *microspores* and larger *megaspores*. The microspores give rise to extremely reduced male prothalli, each bearing an antheridium containing a small number of spermatozoids. The megaspores, which may be reduced to one per sporangium, produce slightly larger female prothalli, each bearing a few archegonia. During their development these gametophytes are retained within the spores. In some species with a single, large megaspore per sporangium, development of gametophyte, fertilization of egg and embryo formation takes place while the megaspore is enclosed within the sporangium which maintains its physiological connexion with the sporophyte. This condition foreshadows that in seed plants where the megaspore is permanently retained within the megasporangium, on the sporophyte. Living representatives are in four orders, Filicales (ferns), Equisetales (horsetails), Lycopodiales (clubmosses), Psilotales. Fossil forms include orders Psilophytales and Sphenophyllales.

PTERIDOSPERMAE. See *Cyadofilicales.*

PTERODACTYLA (PTEROSAURIA). Fossil order of flying reptiles. Lived during Jurassic and Cretaceous. Membraneous wings supported mainly by greatly elongated fourth finger, and also by rest of arm. Many analogies in structure with birds, but independently developed, though both birds and pterodactyls evolved from the same group of non-flying reptiles.

PTEROPODA. Sea butterflies; a few families of gastropod molluscs, highly modified for pelagic life, with or without a shell.

PTERYGOTA (METABOLA). Sub-class of insects including all except a few primitively wingless forms (Apterygota). Some members of the Pterygota are wingless (e.g. fleas) but are believed to have been derived from winged ancestors. Cf. *Apterygota*, *Endopterygota*, *Exopterygota.*

PTYALIN. An amylase present in saliva of some mammals, including man.

PUBIC SYMPHYSIS. Joint or fusion formed mid-ventrally between pubic bones of two halves of pelvic girdle. In most mammals, in *Archaeopteryx* (but not in living birds except ostrich) and in many reptiles. See *Symphysis.*

PUBIS. Ventral, forward-projecting part of hip girdle (q.v.) of tetrapods. See Fig. 10, p. 216.

PULMONARY. (Adj.). Of the lung (q.v.). *P. artery* of vertebrates, carries de-oxygenated blood from heart (from right ventricle in crocodiles, birds, and mammals) to capillaries of lungs. *P. vein* carries oxygenated blood back to heart (to left auricle in all tetrapods).

PULMONATA. Order of gastropod molluscs, members of which have a lung, e.g. snails, slugs.

PULP-CAVITY. Internal cavity of vertebrate tooth or denticle, open by usually a narrow channel to tissues in which tooth is embedded, containing connective tissue, nerves, and blood-vessels, with odontoblasts lining the dentine wall of the cavity. *Persistent pulp*, pulp-cavity widely open, tooth growing continuously from below, e.g. rodent incisor.

PULSE. Wave of raised pressure which passes rapidly (much faster than rate of flow of blood) from heart outwards along all the arteries, every time the ventricle discharges its contents of blood into the aorta. The increased pressure dilates the arteries, and this can be felt; the dilation also relieves the pressure somewhat, so that the pulse dies away as it proceeds and has quite disappeared in the capillaries.

PULVILLUS. Last segment of foot of an insect; it has a claw on either side of a pad.

PULVINUS. Localized enlargement of base of leaf-stalk and/or of base of leaflets in certain plants, concerned in movements of leaves or leaflets in response to stimuli.

PUPA (CHRYSALIS). Stage between larva and adult of endopterygote insect, in which locomotion and feeding cease but great developmental changes occur.

PUPIL. Opening in iris (q.v.) at front of eye. See Fig. 3, p. 86.

PURE LINE. Succession of generations of organisms homozygous for all genes. All variation (except for occasional mutation) in a pure line is due to environmentally produced modifications, which are not inherited. Selection of certain of these modifications for breeding does not therefore alter the characteristics of the pure line; it breeds true. Pure lines are obtained by intensive inbreeding, most efficiently by self-fertilization.

PYCNIDIUM. Minute, globose or flask-shaped fungus fruit body, with apical ostiole, lined internally with conidiophores.

PYCNOSIS. Contraction of nucleus into compact strongly-staining mass, occurring as cell dies.

PYLORUS. Junction between stomach and duodenum of vertebrate. Has a sphincter muscle within a fold of mucous membrane which closes off the junction during digestion of food in stomach.

PYRENOID. Small, round protein granule surrounded by a starch sheath found singly or in numbers embedded in chloroplasts of various Algae.

PYRIDOXINE (VITAMIN B6). Vitamin of B group, forming a coenzyme, required by variety of organisms, e.g. some of the yeasts, bacteria, insects, birds, and mammals; not certainly by man.

PYXIDIUM. Type of capsule (q.v.).

Q

Q_{10}. Temperature coefficient. The increase in rate of a process (expressed as a multiple of initial rate) produced by raising temperature 10°C. Often between two and three for biological, as for many chemical processes.

Q_{O_2}. Term used in measurement of respiration. Oxygen uptake in microlitres (µl.) per milligram dry weight per hour.

QUADRAT. Square of vegetation (standard size is one metre square) chosen at random for study of composition of vegetation in a selected area.

QUADRATE. Cartilage-bone of posterior end of vertebrate upper jaw (developing in palato-quadrate cartilage), attached to neurocranium (brain case), and in most vertebrates articulating with lower jaw. In mammals becomes incus (q.v.).

QUATERNARY. Geological period comprising both Pleistocene (q.v.) and Recent.

R

RACEME. Kind of inflorescence (q.v.).

RACHIS. (1) Main axis of an inflorescence. (2) Axis of a pinnately compound leaf to which leaflets are attached.

RADIALLY SYMMETRICAL. Capable of being halved in either of two (or more) planes so that the halves are approximately mirror images of each other. Characteristic of many animals, e.g. coelenterates, echinoderms; and of many flowers. See *Actinomorphic*, *Bilaterally Symmetrical*.

RADICAL. (Bot.). Arising from the root or crown of the plant.

RADICLE. Root of embryo of seed plants.

RADIOLARIA. Group of marine planktonic rhizopod Protozoa. Their skeletons of silica have been important in forming flint, and oozes of the ocean bottom.

RADIUS. The anterior of the two bones (other is ulna) of fore-arm of tetrapod vertebrate fore-limb. Articulates with side of hand (forefoot) on which thumb is. See Fig. 7, p. 175.

RADULA. 'Tongue' of molluscs; a horny strip with teeth on its surface used for rasping food. It is absent in lamellibranchs, which feed on very small particles of food.

RAMENTA. Brown, chaffy, scaly epidermal hairs covering young leaves and stems of ferns.

RAMET. An independent individual of a clone (q.v.).

RAPHE. (1) In seeds formed from anatropous (q.v.) ovules, longitudinal ridge marking position of adherent funicle (q.v.). (2) In Diatoms, longitudinal slit in valve (q.v.), varying in form and structure; associated with power of movement of these organisms.

RAPHIDE. Needle-shaped crystal of calcium oxalate occurring in bundles in certain plant cells.

RATITES. Living flightless birds. Cf. *Carinates*. Wings reduced in size, without barbules on feathers, or keel on sternum for attachment of wing muscles. E.g. emu, cassowary, kiwi, ostrich, rhea. Term is not now usually used in classification of birds.

RAUNKIAER'S LIFE FORMS. System of classification of vegetation based on position of perennating buds in relation to soil level. Indicates how plants pass the unfavourable season of their annual cycle. The following classes are recognized. *Phanerophytes:* woody plants with perennating buds borne more than 25 cm. above soil level. Includes trees and many shrubs. Taller trees, more than 8 m. high are *Mega-* and *Meso-phanerophytes*; trees and shrubs between 2 and 8 m. are *Microphanerophytes*, and shrubs between 25 cm. and 2 m. are *Nanophanerophytes. Chamaeophytes*: woody or herbaceous plants with perennating buds above soil level but below 25 cm. *Hemicryptophytes*: herbs with perennating buds at soil level, protected by soil itself, or by dry, dead portions of the plant. *Geophytes*: herbs with perennating buds below soil surface. *Helophytes*: herbs with perennating buds lying in mud. *Hydrophytes*: herbs with perennating buds lying in water. *Therophytes*: herbs which survive the unfavourable season as seeds.

REACTION TIME. See *Latent period.*

RECAPITULATION. Occurrence in the embryonic development of an individual, of stages repeating the structure of ancestral adult forms, successively later embryonic stages corresponding to successively more recent ancestors. Hence 'ontogeny repeats phylogeny' (though in condensed form, and with special embryonic adaptations). E.g. the gill-slits of an embryo bird repeat the gill slits of its fish ancestors. Though historically important, recapitulation in this sense is largely discredited now; but it is probably true that the earlier embryonic stages of an organism resemble the corresponding *embryonic* stages of its ancestors, more than do the later stages, and recapitulation of ancestral embryonic characters, in this sense, occurs. Thus the gill slits of a bird embryo resemble the embryonic gill slits of its fish ancestors, the developmental mechanism for this region of the body having changed relatively little in the course of evolution.

RECENT. Holocene (q.v.).

RECEPTACLE. (1) (*Thalamus, Torus*) Apex of flower-stalk bearing flower parts (perianth, stamens, carpels); axis of flower. Variously shaped in different flowers, conical to concave, and according to its form the gynoecium may be *superior* or *inferior* and the flower *hypogynous*, *perigynous*, or *epigynous*. When the carpels are at apex of a conical receptacle and other flower parts inserted in turn below level of carpels the gynoecium is *superior* and the flower *hypogynous*, e.g. buttercup, Fig. 9a. When carpels are at apex (centre) of a concave receptacle with other flower parts borne around its margin, the gynoecium is *superior* and the flower

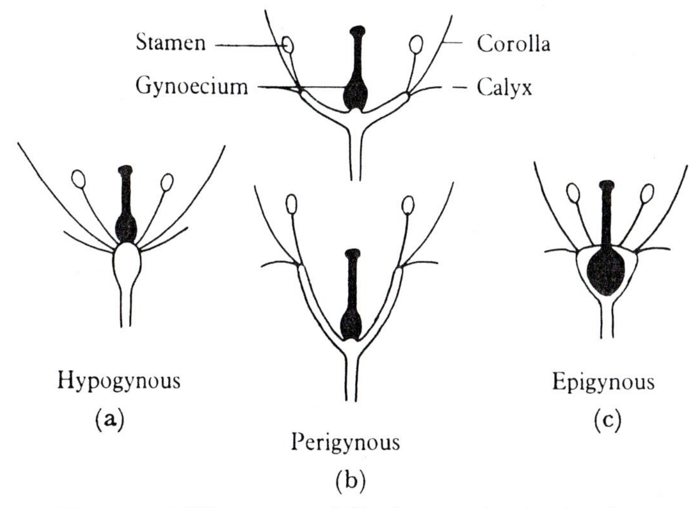

Fig. 9. Diagram of different types of floral receptacle, showing the position of the gynoecium (black) relative to the other flower parts.

perigynous, e.g. bramble, rose. Fig. 9b (upper and lower figures). When receptacle completely encloses carpels and other flower parts arise from receptacle above, the gynoecium is *inferior* and the flower *epigynous*, e.g. apple, dandelion. Fig. 9c. In this condition the carpel walls are intimately fused with wall of the receptacle. (2) Also used to describe the shortened axis of the inflorescence (q.v.) (capitulum) in Compositae.

RECEPTACULUM SEMINIS. Spermatheca (q.v.).

RECEPTOR (SENSE-ORGAN)). That part of an animal which, in co-operation with the nervous system, detects what goes on (i.e. receives stimuli from) outside or inside the animal. Of different kinds, each especially sensitive to a specific sort of stimulus, such as temperature, light, or muscular movement. When a stimulus of sufficient intensity affects an appropriate receptor, the latter influences the behaviour of the animal by sending impulses through the nervous system, nerve-cell(s) of which connect with it.

RECESSIVE. (Genetics). Converse of dominant (q.v.). A recessive gene has no effect on phenotype unless homozygous.

RECIPROCAL CROSSES. Crosses between two plants in which the source of male and female gametes are reversed.

RECOMBINATION. Formation in offspring of combination of genes not present in either parent, by independent assortment (q.v.) of chromosomes and their genes during gamete production followed by random union of different sorts of gamete at fertilization.

RECTUM. Terminal part of intestine, opening to exterior by anus or cloacal opening. In vertebrates, rather narrower than colon (q.v.).

RECTUS MUSCLES OF EYE. See *Eye-muscles*.

RED BLOOD CELLS OR CORPUSCLES (ERYTHROCYTES). Cells of vertebrate blood, containing haemoglobin (q.v.). Carry nearly all the oxygen contained in the blood. Smooth-surfaced, flattened discs, round in cyclostomes and most mammals, oval in other vertebrates. Easily distorted but elastic. Incapable of movement on their own. In mammals (but not in other vertebrates) have no nuclei, except in embryo. An individual r.b.c. has a relatively short life (average about four months in man) and r.b.c.s are formed (in myeloid tissue of bone marrow) and perish (the remains being taken up by reticulo-endothelial system, q.v.) continuously in enormous numbers (over a million a second in man). In man each is a circular biconcave disc, eight microns in diameter after fixation, and there are roughly five million per cubic millimetre of blood.

REDIA. One of the kinds of larva produced asexually by a previous larval stage of flukes; parasitic in snails; rediae reproduce asexually giving rise either to more rediae or to cercariae (q.v.).

REDUCTION DIVISION. Meiosis (q.v.).

REFLEX. Very simple form of behaviour occurring in almost all animals with a nervous system, in which a certain kind of stimulus almost invariably evokes, with hardly perceptible delay, one specific kind of simple response. E.g. a pin stuck in one's foot evokes an immediate withdrawal. The constancy and immediacy of response depends on an inborn nervous pathway (i.e. one which is independent of experience), the *reflex arc*, along which the impulses travel. It involves the central nervous system. In the instance cited the stimulus causes groups of impulses to travel along a number of sensory nerve fibres into the spinal cord, where they are relayed through other nerve fibres, and impulses are set going down a group of motor nerve fibres which activate the muscles whose contraction withdraws the foot. Even the simplest reflex involves synapses in the central nervous system, and these provide opportunity for interaction of the nerve-cells involved in the reflex with other events in the central nervous system, allowing for example development of conditioned reflexes (q.v.), or inhibition (q.v.).

REFUGIUM. Locality which has escaped drastic alteration following climatic change, in contrast to the region as a whole, e.g. a driftless area (usually a tableland or mountain, known as a *Nunatak*) within a region that has undergone general glaciation. Usually forms a centre for relic (q.v.) species. Exist widely over earth's surface because although, owing to glacial advance and recession, changes of climate affecting the survival of plants have occurred, no general climatic changes affects a large area uniformly, with the result that relic colonies, of species or communities, are left behind.

REGENERATION. Restoration by an organism of tissues or organs which have been removed. Power of regeneration is highly variable between different groups of organisms; in some Platyhelminthes a minute fragment can regenerate the whole animal; in mammals limited to wound-healing, regrowth of peripheral nerve fibres, and compensatory hypertrophy (q.v.).

REGULATION. Normal development of an animal embryo, or part of an embryo, occurring although its structure has been disturbed in some way, as by removing some of it, or adding to it, or rearranging it. E.g. half a sea-urchin embryo, at a sufficiently early stage, will develop into a normally proportioned though small larva. Regulation is a widespread feature of the early stages of animal development. In some species powers of regulation practically disappear at the time of fertilization (*mosaic eggs*, q.v.) while in others the fertilized egg, and later stages, easily regulate (*regulation eggs*).

RELEASER. Stimulus which sets off an instinctive act. A *social releaser* is a releaser emanating from a member of the same species as the reactor.

RELIC. Surviving organism, population, or community, characteristic of an earlier time. See *Refugium*.

RELIC (OR RELICT) DISTRIBUTION (R. FAUNA, R. FLORA). Spatial distribution which represents localized remains of an originally much wider distribution. E.g. organisms widespread in glacial times now confined to mountain tops. See *Relic*, *Refugium*.

RENAL PORTAL SYSTEM. In fish and Amphibia blood is brought from capillaries of posterior end of body (hind limbs, tail) to kidneys by a pair of renal portal veins really part of posterior cardinal veins), which break up into capillaries mainly around kidney tubules. Blood is collected up again either into posterior cardinal veins (fish) or renal veins and vena cava inferior (Amphibia). This portal system (q.v.) is slightly or not at all developed in amniotes, which have a different sort of kidney (metanephros, q.v.).

RENNIN. Enzyme secreted in the stomach of young mammals which clots milk (converting soluble protein caseinogen into casein which forms insoluble calcium-casein compound). Rennet is impure rennin.

REPLACING BONE. Cartilage-bone (q.v.).

REPLICATION. Production of exact copies of complex molecules, as must occur during growth of any living thing.

REPTILIA. Reptiles, a class of vertebrates. Includes many orders, the living ones being Chelonia (turtles and tortoises), Rhyncocephalia (the tuatara), Squamata (lizards and snakes), Crocodilia (crocodiles); and the entirely extinct ones being, amongst others, Ornithischia and Saurischia (both dinosaurs), Pterodactyla, Ichthyosauria, Plesiosauria, Therapsida (the ancestors of the mammals).

Reptiles are poikilothermic (q.v.) tetrapods, mostly living and reproducing entirely on land, though some have become aquatic. They were the dominant terrestrial animals in the Mesozoic. Modern ones are distinguished from Amphibia by horny skin, metanephric kidney, and by the embryo being enclosed in an egg-shell and having a protective amnion and an allantois; and from birds and mammals by absence of the special features, such as feathers or hair, of those groups. But in the fossil record there are intermediates between them and both amphibia and mammals, and no sharp dividing line is possible. The distinction for most fossils is made on the structure of the vertebral column, and other skeletal features.

RESOLUTION. See *Microscope*.

RESPIRATION. (1) Breathing, e.g. pumping air in and out of lungs, or water over gills. (2) Taking oxygen from the environment and giving off carbon dioxide. (1) and (2) are sometimes referred to as *external respiration* in contrast to (3) *internal, tissue* or *cell respiration*, i.e. the chemical reactions from which an organism derives energy. In most organisms internal respiration is accompanied by consumption of free oxygen and production of carbon dioxide (the external manifestation of which is (2); in some organisms (2) being facilitated by (1)). This kind of respiration is called *aerobic*. If glucose, for instance, is being used as basis for aerobic respiration, it is oxidized to carbon dioxide and water, and the same amount of energy is liberated as if that glucose were burnt in air. In the organism a complicated chain of oxidation-reduction reactions is involved, various substances being alternately oxidized and reduced under the influence of various enzyme systems (see *Cytochrome, Dehydrogenase, Oxidase*). Energy can be liberated by the break-down of substances without molecular oxygen being concerned in the reaction; this is known as *anaerobic respiration*, e.g. breakdown of glycogen to lactic acid in vertebrate muscle; or of glucose to ethyl alcohol and carbon dioxide by yeast. Such processes do not liberate as much energy as does aerobic respiration. Many organisms (or parts of them) respire anaerobically for some time when their supply of oxygen is insufficient for aerobic respiration, in which case they may acquire an oxygen debt (q.v.). A few bacteria are strictly anaerobic, i.e. they do not use free oxygen at all.

RESPIRATORY ENZYME. Enzyme which catalyses oxidation-reduction reactions. See *Dehydrogenase, Oxidase*.

RESPIRATORY MOVEMENT. Movement of part of an animal by which the air or water from which the animal obtains oxygen and to which it gives off carbon dioxide is frequently renewed at the surfaces where this exchange of respiratory gases occurs (i.e. at the respiratory organ). There is great variety of respiratory organs among animals, and great variety of respiratory movement. E.g.

breathing movements of chest and diaphragm in man and other mammals; telescopic movements of the abdomen of a wasp; baling of water past the gills in a lobster.

RESPIRATORY ORGAN. Organ across whose surface interchange of respiratory gases (carbon dioxide and oxygen) occurs, between body fluids of animal and the environment; lung, gill.

RESPIRATORY PIGMENT. Substance which combines reversibly with oxygen, thus acting as a carrier or store of it. E.g. haemoglobin of human blood becomes loaded with oxygen in the lungs, where it comes into equilibrium with air; and it gives up this oxygen when it comes into contact with tissues having a low oxygen pressure. Respiratory pigments change colour according to the degree of oxygenation (e.g. haemoglobin is scarlet oxygenated, purple deoxygenated) and many have characteristic absorption spectra; these properties have facilitated their investigation. Respiratory pigments are frequently present in blood, either in plasma (haemocyanin, chlorocruorin, haemoglobin of several invertebrates) or in blood corpuscles (haemoglobin of vertebrates; a few other pigments of invertebrates).

RESPIRATORY QUOTIENT (R. Q.). Ratio of the volume of carbon dioxide expired to the volume of oxygen consumed during the same time. A theoretical R.Q. can be calculated for the oxidation of various foodstuffs. For carbohydrates it is 1, for fats 0.7, and for proteins 0.8. It was formerly thought that the R.Q. of an organism or tissue gave direct information concerning the kind of foodstuff being oxidized; but in practice the interpretation of an R.Q. is not so simple, for cells can oxidize more than one of the three foodstuffs at the same time, and certain other metabolic processes may use oxygen or produce carbon dioxide.

RESPONSE. Change in an organism or part of it, usually adaptive, produced by a stimulus (q.v.). See *Irritability*.

RESTING CELL (R. NUCLEUS). Cell or nucleus which is not undergoing mitosis (is in *resting stage*). May be very active in other ways.

RESTING EGGS (WINTER EGGS). Eggs which can only develop after a dormant period, during which they are resistant to adverse conditions; produced by some fresh water animals, e.g. rotifers, certain crustacea (water fleas); may differ considerably from other eggs of the same species, in, e.g. being fertilized instead of parthenogenetic, having more yolk, or thicker shell.

RESTING POTENTIAL. See *Action Potential*.

RETICULATE (THICKENING). Internal thickening of wall of a xylem vessel or tracheid in the form of a network.

RETICULIN FIBRES. Very fine, almost inextensible intercellular fibres forming a network around and amongst the cells in many vertebrate tissues, e.g. muscle, nerve, kidney, liver, and other glands. They hold the tissues together. They stain selectively with special silver impregnation methods (hence name *argyrophil fibres*) and they are more resistant to chemical treatment and high temperature than collagen fibres. Nevertheless their material (*reticulin*) seems very closely related to collagen, and transitional states are frequently found. Indeed in formation of collagen fibres (in embryo or after wounding), reticulin fibres appear first and change to collagen.

RETICULO-ENDOTHELIAL SYSTEM. Those phagocytic macrophages of vertebrates in contact with blood (in bone-marrow, spleen, liver) or lymph (in lymph nodes); they free these fluids from foreign particles, e.g. bacteria. See *Macrophage*.

RETINA. Layer lining interior of vertebrate eye, except at front (in region of ciliary body); it is sensitive to light. It has an outer pigmented layer, next to the choroid; and an inner transparent nervous layer next to cavity of eye-ball. The nervous layer contains light-sensitive rods (q.v.) and/or cones (q.v.) in contact with pigmented layer; and nerve-fibres, intermediary nerve-cells, blood-vessels and glia, interposed between incoming light and rods and cones. Embryologically part of the brain. Develops as hollow outpushing of brain wall, whose outer end becomes dinted in to form a stalked cup; the double-walled cup forms the two layers of the retina, the stalk forms optic nerve. See Fig. 3, p. 86; *Fovea*; *Blind spot*. In molluscs the retina is formed not from the nervous system but from the external ectoderm, and the light-sensitive cells are external to the nerve-fibres, etc.

RH FACTOR (RHESUS FACTOR). Substance (an antigen; there are actually several, closely related) occurring in blood corpuscles of a high proportion of human beings (85 per cent in Great Britain and U.S.A.). Depends on a complex of linked genes. The rest of the population is without the factor (Rh-negative). Rh-negative individuals do not normally possess antibodies against the antigen they have not got (as they do in the classical blood-groups, q.v.). But they can acquire such antibodies, as by blood transfusion; or in the case of a woman by bearing Rh-positive children, whose antigen crosses the placenta. The antibody thus formed in a Rh-negative mother may cross the placenta again and if in sufficient concentration, which usually requires at least one preceding pregnancy with Rh-positive child, it may damage a Rh-positive foetus.

RHIZOID. Single or several-celled hair-like structure serving as a root. Rhizoids are present at base of moss stems and on under-surfaces of liverwort plants and fern prothalli.

RHIZOME. Underground stem, bearing buds in axils of reduced scale-like leaves; serving as a means of perennation and vegetative propagation, e.g. mint, couch grass.

RHIZOMORPH. Compact strand of fungus hyphae, capable of increase in length by apical growth, that serves for transport of food materials from one part of thallus to another and assists in spread of fungus through or over substratum.

RHIZOPODA (SARCODINA). Class of Protozoa characterized by the possession of pseudopodia; includes *Amoeba*; Foraminifera; Radiolaria, etc.

RHIZOSPHERE. Zone immediately surrounding roots, characterized by enhanced microbiological activity.

RHODOPHYCEAE. Red algae (seaweeds). Class of Algae with chromoplasts containing phycoerythrin and phycocyanin in addition to chlorophyll. Usually red in colour because of predominance of phycoerythrin. Widely distributed marine algae. Thallus filamentous, ribbon-shaped or more laterally expanded, with relatively complex internal structure. Reproduction unique in that motile reproductive cells are absent.

RHYNCHOCEPHALIA. A small order of reptiles, containing only *Sphenodon* (the tuatara) and a number of extinct forms. Lizard-like but with many primitive features.

RHYNCHOTA. Hemiptera (q.v.).

RHYTIDOME. See *Bark*.

RIBOFLAVIN (= LACTOFLAVIN, VITAMIN B_2). Vitamin of B group. Forms part of various enzymes concerned in cellular oxidation, widely (universally?) distributed in living organisms. Liver, muscle, yeast are important sources. Required as vitamin by those vertebrates (including man) and insects tested, and by some bacteria. Synthesized by many bacteria, including those present in human intestine.

RIBOSENUCLEIC ACID (RNA). See *Nucleic Acid*.

RIBOSOMES. Cytoplasmic granules, about 100 A in diameter and hence invisible with a light microscope, very rich in RNA, concerned in protein synthesis, often attached to endoplasmic reticulum (q.v.).

RINGER (RINGER'S FLUID). Physiological saline (q.v.) containing sodium, potassium, and calcium chlorides (and sometimes other salts), these being main salts in fluid which normally bathes cells. Much used in physiological experiments for temporarily maintaining cells or organs alive in vitro (q.v.).

RNA. Ribosenucleic acid. See *Nucleic Acid*.

ROD. Kind of light-sensitive nerve-cell present in vertebrate retina. Rods do not discriminate fine detail or probably colour differences, but they are sensitive to very dim light, which cones (q.v.) are not. There are about 120 million in one retina of a primate. Response to light depends on visual purple (q.v.).

RODENT. Any member of the order Rodentia, e.g. rat, squirrel (rabbit and hare are now often put in a separate order, Lagomorpha). Gnawing mammal, with a pair of large continually growing chisel-like incisors in upper and lower jaws. Grinding molars. Most widespread and numerous of all mammal orders.

ROGUE. (1) Variation from standard type of variety, etc.; (2) to remove such variants from a growing crop; (3) used also in a wider sense meaning to remove any undesirable plants from a variety in the field, including genetic variants, those affected by disease, etc.

ROOT. That part of vascular plants that usually grows downwards into the soil, anchoring plant and absorbing water and nutrient salts. Roots cannot be distinguished from stems on basis of their position with respect to soil since roots of some plants may be wholly or partly above ground, while other plants possess underground stems. Externally, roots differ from stems principally in that they do not bear leaves and buds; they also possess at their tip a protective layer of cells, the root-cap. Internally, roots differ from stems in arrangement of vascular tissue in a central core and with protoxylem (q.v.) exarch.

ROOT-CAP. Cap of loosely arranged cells covering apex of growing point of root and protecting it as it is forced through soil. Formed from promeristem (q.v.), dermatogen (q.v.), or from a meristematic layer external to dermatogen known as *calyptrogen.*

ROOT HAIR. Tubular outgrowth of epidermal cell of root; with thin delicate wall having intimate contact with soil particles. Produced in large numbers behind root tip forming piliferous layer and greatly increasing absorbing surface of root.

ROOT NODULES. Small swellings on roots of leguminous plants (pea, bean, clover, etc.) produced as a result of infection by nitrogen-fixing bacteria. See *Nitrogen Fixation.*

ROOT PRESSURE. Pressure under which water passes from living cells of root into xylem; demonstrated by exudation of liquid from cut end of decapitated plant.

ROTIFERA. Wheel animalcules. Phylum of minute Metazoa which swim and feed by means of a ciliated band, the 'wheel'. Formerly confused with ciliate Protozoa and included in Infusoria.

ROUNDWORMS. Nematoda (q.v.).

R. Q. Respiratory quotient (q.v.).

RUDERAL. Plant living in waste places near habitations.

RUMEN. Storage compartment of complicated stomach (rumen is really a diverticulum of oesophagus) of ruminants, into which newly eaten, but unchewed, food is passed, and from which it (the 'cud') is subsequently returned to mouth for chewing. Some digestion of food (especially cellulose), and much synthesis of B vitamins which are absorbed by the animal, occurs by bacterial action in rumen; some products of cellulose digestion are absorbed by rumen.

RUMINANT. Mammal belonging to the sub-order Pecora of the order Artiodactyla. Deer, giraffes, sheep, goats, antelopes, oxen. No upper incisor teeth. Often with bone-cored horns. Stomach usually complicated. See *Rumen.*

RUNNER. Stolon that roots at tip forming new plant that eventually is freed from connexion with parent by decay of runner. Also used horticulturally of the daughter plant itself.

S

SACCHARASE. See *Sucrose.*

SACCHAROMYCES. Genus of ascomycete Fungi that includes those used in production of bread and alcoholic drinks. See *Yeasts.*

SACCULE. See *Ear, inner.*

SACRAL VERTEBRA. Vertebra of tetrapods which articulates by means of rudimentary ribs with ilia of hip-girdle (q.v.). One in Amphibia, two or more fused together in other tetrapods.

SACRUM. Group of sacral vertebrae fused together, the ilia of the hip-girdle being united to some or all of them. See *Hip-Girdle.*

SAGITTAL. In the plane extending longitudinally (antero-posteriorly), and dorso-ventrally in the mid-line. Divides bilaterally symmetrical animal into two similar (right and left) halves.

SALIVA. Fluid secreted into mouth. In terrestrial vertebrates, fluid containing mucus (q.v.), and in some of them ptyalin (q.v.), secreted from salivary glands into mouth as a reflex response to presence there of food. Provides lubrication for swallowing. In insects, fluid secreted on to mouth-parts, often containing digestive enzymes; used for dissolving or lubricating food, and in blood-suckers may contain an anti-coagulant.

SALIVARY GLAND CHROMOSOMES (SALIVARIES). Giant chromosomes occurring in salivary glands (and some other tissues) of dipterous insects (including *Drosophila*). Nuclei of these tissues have their chromosomes microscopically visible, unlike normal resting nuclei. Each pair of homologous chromosomes is closely adherent together (paired). The chromosomes are almost uncoiled, stretched out to practically their full length (which may be $\frac{1}{5}$ mm.) and greatly thickened by repeated duplication. They are marked by an elaborate pattern of transverse basophilic bands, formed by homologous chromomeres of the numerous chromosome strands lying side by side. The pattern is due to the arrangement of the genes along the chromosomes. From genetic differences associated with changes in the pattern, numerous genes have been localized. See *Chromosome Map.*

SALTATION. Abrupt, often permanent change in character of whole, or more usually, part, of fungus thallus. In culture, generally appearing as a sector differing, e.g. in rate of growth, colour, degree of spore production. May be due to mutation (q.v.) or to somatic segregation from a heterokaryon (q.v.) of nuclei of unlike genetical constitution.

SAMARA. Winged achene (q.v.).

SAPROPHYTE. Organism which obtains organic matter in solution from dead and decaying tissues of plants or animals. Cf. *Parasite.*

SAPWOOD. Outer region of xylem of tree trunks, containing living cells, e.g. xylem parenchyma and medullary ray cells, and functioning in water conduction and food storage as well as providing mechanical support. Cf. *Heartwood.*

SARCODINA. Rhizopoda (q.v.).

SAURIA. Lacertilia (q.v.).

SAURISCHIA. An order of fossil reptiles, one of the two into which the group Dinosauria has now been split (the other being Ornithischia). Lived during the Mesozoic. Contained carnivorous and herbivorous species. Many bipedal. Some, but by no means all, were extremely large (up to fifty tons).

SAUROPSIDA. Term not usually used in classification but denoting a natural group: the majority of reptiles including all the living forms and many fossil forms, e.g. dinosaurs, pterodactyls, and the birds, all of which are closely related; as distinct from the extinct mammal-like reptiles (Therapsida, q.v.) and the mammals, which form another natural group (Theropsida).

SCALARIFORM THICKENING. Internal thickening of wall of a xylem vessel or tracheid, in the form of more or less transverse bars, suggestive of ladder rungs.

SCAPE. Leafless flowering stem arising from ground level, e.g. dandelion, daffodil.

SCAPHOPODA. Elephant's tusk shells; small class of molluscs. Marine; specialized for burrowing; tubular shell open at both ends.

SCAPULA. Dorsal part of shoulder girdle (q.v.). Shoulder-blade in man and other mammals. See Fig. 10, p. 216.

SCHISTOSOMA (BILHARZIA). Genus of blood-flukes (of class Trematoda, Platyhelminthes) including human parasites of very great medical importance, in e.g. parts of Africa (especially Egypt), South America, China.

SCHIZOCARP. Dry fruit formed from a syncarpous ovary that splits at maturity into its constituent carpels forming a number of partial fruits, usually one-seeded, resembling achenes (q.v.) and called mericarps, e.g. hollyhock, mallow, geranium, cow parsley.

SCHIZOGENOUS. (Of secretory cavities in plants), originating by separation of cells, e.g. oil-containing cavities in leaves of St John's wort. Cf. *Lysigenous.*

SCHWANN CELL. Kind of cell which ensheaths every nerve-fibre of vertebrate peripheral nervous system. In a myelinated fibre one cell occurs between every pair of adjacent nodes (q.v.). Probably largely derived from neural crest.

SCION. Twig or portion of a twig of one plant that is grafted on to a stock of another.

SCLEREID. See *Sclerenchyma.*

SCLERENCHYMA. Tissue giving mechanical support to plants. Cells thick walled, usually lignified, inelastic, sometimes consisting of cellulose, often so thick as to leave only a very small lumen; usually without living protoplasm at maturity. Two types of cell occur in sclerenchyma, *fibres* and *stone-cells* (*sclereids*). Fibres are very elongated cells with tapering ends, occurring singly or variously grouped into strands. They are extracted from such plants as flax (cellulose fibres) and hemp (lignified fibres) and used in manufacture of rope, linen, paper, etc. Stone cells are usually not much longer than they are broad, occurring singly or in groups; common in fruits, e.g. pear, nuts, and in seed coats.

SCLEROPROTEINS. Very stable fibrous proteins, insoluble in water or salt solutions; present mainly as surface covering of animals, e.g. keratin, collagen, or as fibres binding cells together, e.g. collagen, elastin.

SCLEROSIS. Hardening. E.g. in vertebrate tissues after injury, by increase in concentration of collagen.

SCLEROTIC. Fibrous or cartilaginous firm outer coat of eye-ball in vertebrates, continuous with cornea in front of eye. See Fig. 3, p. 86.

SCLEROTIUM. Compact tissue-like mass of fungus hyphae, often with a thickened rind, varying in size from that of a pin's head to that of a man's head; capable of remaining dormant for long periods, e.g. during conditions unfavourable to normal fungus growth, and commonly giving rise to fruit bodies, e.g. ergot (q.v.).

SCOLEX. Part of tapeworm which is attached to wall of gut of host by suckers and/or hooks; sometimes called the 'head'.

SCROTUM. Pouch of skin of pelvic region in many mammals which contains testes (at least during breeding season) keeping them cooler than body temperature (latter inhibits development of sperm).

SCUTELLUM. Cotyledon of grass embryo.

SCYPHISTOMA. The fixed (polyp) stage of jelly-fish (scyphozan Coelenterata), usually quite small and insignificant, which by horizontal splitting gives rise to free-swimming jelly-fish.

SCYPHOMEDUSAE. Scyphozoa (q.v.).

SCYPHOZOA (SCYPHOMEDUSAE). Jelly-fish. Class of coelenterates (of subphylum Cnidaria). Polyp stage (scyphistoma, q.v.) small and inconspicuous or absent; medusae large and more complexly organized than those of Hydrozoa; gonads endodermal. Cf. *Hydrozoa*, *Actinozoa*.

SEAWEEDS. Red, brown, or green Algae (q.v.), living in or by the sea.

SEBACEOUS GLAND. Skin gland of mammals nearly always opening into a hair-follicle, secreting fatty substance (*sebum*). Formed from epidermis, but projects deep into dermis.

SECONDARY MERISTEM. Region of active cell division that has arisen from permanent tissue, e.g. cork cambium (phellogen), wound cambium. Cf. *Primary meristem*.

SECONDARY SEXUAL CHARACTER. A characteristic of animals which differs between the two sexes; but excluding the gonads, and the ducts, with their associated glands, which convey the gametes.

SECONDARY THICKENING. Formation of additional, secondary, vascular tissue by activity of cambium (q.v.) with accompanying increase in diameter of stems and roots of dicotyledons and gymnosperms; providing additional conducting and supporting tissue for growing plant and, in most cases, making up greater part of mature structure.

SECRETIN. Hormone which stimulates secretion of digestive juices by pancreas, and of bile by liver. Produced in epithelium of wall of duodenum and jejunum under stimulus of acid products of digestion coming through pylorus from stomach. A peptide.

SECRETION. (1) The passage of (usually complex) material elaborated by a cell from the inside to the outside of its plasma membrane; the material (itself called a secretion) having a special function in the organism. Secretion is probably an activity of most cells, but is specialized in gland-cells. Usually takes place by extrusion of material, but may involve destruction of whole cell (holocrine secretion, e.g. sebaceous gland). (2) Secretion, when used in contrast to simple diffusion, means that the cell does work in passing the secretion through its plasma-membrane against the forces of diffusion, e.g. secretion of hypotonic fluid such as sweat. See *Active Transport.*

SECTION. See *Microtome.*

SEED. Product of a fertilized ovule, consisting of an embryo enclosed by protective seed coat(s) derived from the integument(s). Characteristic structure of gymnosperms and angiosperms (seed plants). In some (*endospermic*) seeds embryo is surrounded by nutritive *endosperm* tissue, e.g. castor oil, pine; in other (*non-endospermic*) seeds endosperm is absent, e.g. pea. See *Endosperm.*

SEGMENTATION. (1) (Zool.). *Metameric segmentation. Metamerism.* Repetition of a pattern of elements belonging to each of the main organ systems of the body, along the antero-posterior axis of the body; or a comparable repetition along the axis of an appendage. Most strongly marked in Annelida and Arthropoda. The repetition produces a series of units, called *segments* (or *metameres*), of fundamentally similar structure, e.g. in the earthworm each of the externally visible rings is a segment; each segment contains a pattern of blood-vessels, epidermal structures, nervous system, excretory organs, etc., which is repeated with minor variation in every other segment. It should be noted that the similarity between different segments of a given animal may be very imperfect. In particular the segments at the anterior end, which form the head, are very different from each other and from other segments. Vertebrates show segmentation most clearly in embryonic development, but it is almost confined to parts of the muscular, skeletal, and nervous systems. See *Somitic Mesoderm.* It does not appear in the epidermis, as in Annelida and Arthropoda. (2) In embryology, a synonym for cleavage (q.v.) applied to both plants and animals.

SEGREGATION. Separation into different gametes, and thence into different offspring, of the two members of any pair of allelomorphs possessed by an individual; the two allelomorphs having previously been brought together in the individual, one from each of its parents; and having in no way blended or altered each other while associated together in the individual. What is known as Mendel's first law asserts that allelomorphs segregate. Segregation is the result of the relatively unchanging nature of genes; of their separate carriage in chromosomes; of the mechanism of meiosis which provides each gamete with only one member of each pair of chromosomes; and of fertilization by which pairs of chromosomes are formed again.

SEISMONASTY. (Bot.). Response to a non-directional shock stimulus, e.g. rapid folding of leaflets and drooping of the leaves in *Mimosa pudica* when lightly struck or shaken.

SELACHII. A sub-class or order of Chondrichthyes, containing the sharks, dogfish, skates, and rays. Sharp, rapidly replaced teeth, and upper jaw not fused to skull (cf. *Holocephali*). Occasionally extended to cover various fossil groups of Chondrichthyes also; occasionally restricted solely to sharks and dogfish.

SELECTION PRESSURE. Measure of the effectiveness of natural selection in altering the genetic composition of a population.

SELF-DIFFERENTIATION. Of a region, tissue, or feature of an embryo, differentiation, independently of influence of other parts of embryo. Cf. *Dependent differentiation.*

SELF-FERTILIZATION. Fusion of male and female gametes from same individual. Cf. *Cross-Fertilization.*

SELF-POLLINATION. Transference of pollen from anther to stigma of the same flower, or to stigma of another flower on same plant. Cf. *Cross-Pollination.*

SELF-STERILITY. Of some hermaphrodites (q.v.) inability to form viable offspring by self-fertilization. See *Incompatability.*

SEMEN. Product of male reproductive organs, consisting of sperm together with secretions of various accessory glands, e.g. of prostate gland in mammals.

SEMICIRCULAR CANALS. Semicircular tubes, projecting from and joined at both ends to the inner ear (q.v.) of vertebrates. Gnathostomes have three on each side, occupying the three planes of space, two vertical at right-angles to each other, one horizontal. Agnatha have only the two vertical (in some only one). Angular acceleration (e.g. rapid turning) of the animal in any direction produces currents of fluid (endolymph) along the corresponding canal, which are detected by fine projections from sensory cells, and communicated to brain via auditory nerve.

SEMINAL VESICLE (VESICULA SEMINALIS). Organ which stores sperm in the male.

SEMINIFEROUS TUBULES. Coiled tubes (each about 50 cm. long and

$\frac{1}{5}$ mm. diameter in man), several hundreds of which are present in vertebrate testis. Made of germinal epithelium. During sexual activity all stages of spermatogenesis occur in them, up to almost mature sperm. Supposedly nutritive Sertoli cells, to which developing spermatids are attached, are also present. Correspond to ovarian follicles (q.v.) of ovary. See *Follicle-Stimulating Hormone.*

SENSE-ORGAN. Receptor (q.v.).

SENSORY. Concerned with receptors (q.v.). *S. nerve.* Peripheral nerve consisting of nerve-fibres of sensory nerve-cells. *S. nerve-fibre.* Nerve-fibre of a sensory nerve-cell. *S. nerve-cell* (*S. neurone*). Nerve-cell whose fibres connect with receptor, transmitting impulses started by receptor to central nervous system (q.v.). In vertebrates the cell-body of such a nerve-cell is attached to its nerve-fibre at a point along its course situated in dorsal root ganglion, and there are no dendrites. In invertebrates cell-body is situated at (peripheral) end of nerve-fibre furthest away from central nervous system. *S. Root.* Of vertebrates, nerve-root (q.v.) containing the sensory nerve-fibres.

SEPAL. One of the parts forming calyx of dicotyledonous flowers; usually green and leaf-like. See *Flower.*

SEPTICIDAL. (Bot.). Describing dehiscence of multilocular capsule by longitudinal splitting along septa between carpels, separating carpels from each other, e.g., St John's Wort. Cf. *Loculicidal.*

SEPTUM. Partition or wall.

SERE. Particular example of plant succession (q.v.). Seres originating in water are known as *hydroseres*; those arising under dry conditions as *xeroseres*, of which those developing on exposed rock surfaces are known as *lithoseres.*

SEROUS MEMBRANE. Mesothelium and underlying connective tissue lining coelomic spaces (pericardial, perivisceral, pleural, peritoneal cavities) in vertebrates.

SERTOLI CELLS. See *Germinal Epithelium; Seminiferous Tubule.*

SERUM. Blood serum (q.v.).

SESAMOID BONE. Bone developed within tendon of vertebrate, occurring especially in mammals, particularly where tendon works over ridge of underlying bone, e.g. patella (knee-cap).

SESSILE. (1) Of parts of plants or eyes of Crustacea; without a stalk. (2) Of animals, fixed to the substratum; sedentary.

SETA. (Bot.). Stalk of sporogonium (q.v.) in mosses and liverworts. (Zool.). Bristle or 'hair' of invertebrates. (True hair (q.v.) is a feature of mammals only.) Produced by epidermis, consisting either (*a*) solely of cuticular material, or (*b*) of hollow projection of cuticle enclosing an epidermal cell or part of one. Type (*a*) also known as chaetae (q.v.) are characteristic of Polychaeta and Oligochaeta. Setae of insects are of type (*b*).

SEX-CHROMOSOME. Chromosomes of which there is a homologous pair in the nuclei of one sex (the homogametic sex, q.v.), and a dissimilar pair (or an unpaired chromosome) in those of the other (the heterogametic sex, q.v.). The sex chromosomes which occur as a pair in the homogametic sex are called X-chromosomes. The heterogametic sex has an X-chromosome and either a Y-chromosome or none. The homogametic (or XX) sex produces gametes which are identical in their sets of chromosomes, all containing one X-chromosome. The heterogametic (XY or XO) sex produces equal numbers of two different sorts of gametes, one with and one without an X-chromosome. Union of the gametes of the XX sex with those of the XY (or XO) sex therefore results in offspring consisting of approximately equal numbers of the two sexes. The presence or absence of a Y-chromosome determines sex in mammals; in insects it is determined by the number of X-chromosomes in relation to the number of autosomes. The sex-determining mechanism is of course absent in hermaphrodite animals and plants.

SEX-DETERMINATION. See *Sex-chromosomes.*

SEX-LIMITED CHARACTER OR GENE. Gene expressed in only one of the two sexes, e.g. gene controlling characteristic of milk production.

SEX LINKAGE. Of a gene or character, having special distribution with reference to sex as a result of (in the case of a gene) being carried on the X-chromosome (q.v.) or (in the case of a character) controlled by a gene so carried. A man, for instance, receives his X-chromosome and therefore all his sex-linked genes from his mother; and he hands them on to his daughters (who also get an X-chromosome from their mother), not to his sons. Since in a man there is no partner to the X-chromosome, recessive genes on it cannot be masked by a dominant allelomorph, so that men manifest a larger number of recessive genes (like those for haemophilia or red-green colour-blindness) than women; and conversely a smaller number of dominant genes. What is true for men is true for the heterogametic sex of any organism. Sex-linkage must be distinguished from sex-limitation (q.v.). Both sexes manifest a full set of sex-linked genes; but the homogametic sex, unlike the heterogametic sex, will only manifest a recessive gene if it has received it from both parents, which will be a rare event if the gene is uncommon in the population.

SEX RATIO. Relative proportion of the numbers of males and females in a population. Usually expressed as number of males per 100 females. *Primary* sex ratio is that immediately after fertilization; *secondary*, that at birth (or hatching); *tertiary* at maturity.

SEXUAL REPRODUCTION. Reproduction involving fusion of haploid nuclei, usually of gametes (q.v.).

SEXUAL SELECTION. A possible mode of evolution suggested by Darwin. The females of a population are supposed to select for mating only males having certain characteristics and thereby to transmit these characteristics to future generations to the exclusion of characteristics of rejected suitors. Darwin explained the origin of many secondary sexual characters in this way, especially brilliant adornment and courtship behaviour. In Darwin's strict sense sexual selection is probably of rather limited importance since even where a female has a choice between males, which is infrequent, the rejected males will usually find a mate elesewhere and so their characteristics will not be eliminated. The evolution of secondary sexual characters is probably more of a matter of natural selection (q.v.), one pair (male and female) of animals reproducing more successfully than another because of the characters they possess, including those characters serving sexual stimulation.

SHELLFISH. Fishmonger's term for shelled molluscs and crustaceans.

SHIP-WORM. Wood-boring member of the Lamellibranchiata (genus *Teredo*) which does great damage to wooden wharves, ships, etc.

SHORT-DAY PLANT. See *Photoperiodism.*

SHOULDER GIRDLE (PECTORAL GIRDLE). Skeletal support in body-wall for attachment of front fins or limbs of vertebrate. Consists primitively of a curved bar of cartilage or bone on each side of the body, the two often fusing ventrally, forming a hoop, incomplete dorsally, transverse to long axis. Each bar bears a joint with fin or limb. See *Glenoid Cavity.* Region dorsal to joint is *scapula,* region ventral is *coracoid.* Extra dermal bones, notably *clavicle,* usually occur on ventral side. Scapula forms no joint with or bony attachment to vertebral column or ribs (unlike pelvic girdle). Coracoid and clavicle are joined mid-ventrally to breast bone (sternum) in tetrapods. See Fig. 10.

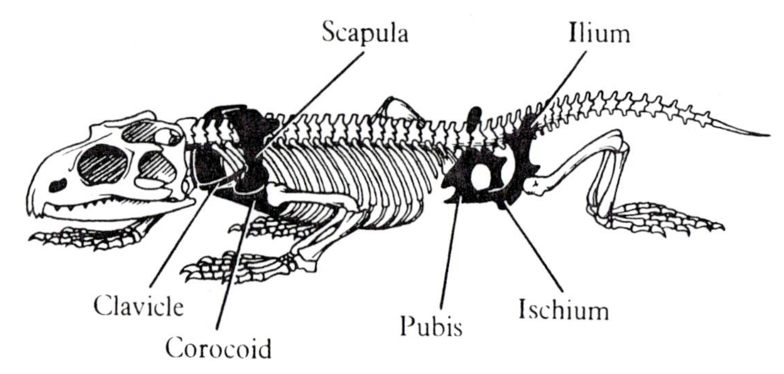

Fig. 10. Skeleton of a tetrapod vertebrate, showing arrangement of shoulder and hip-girdles.

SIBLINGS (SIBS). Brothers and/or sisters; offspring of same male parent and same female parent.

SIEVE-PLATE. See *Sieve-Tube.*

SIEVE-TUBE. Characteristic element of phloem functioning in the transport of food materials synthesized within the plant, e.g.

sugars, proteins. Consists of thin-walled, elongated, living cells arranged in a longitudinal row and forming a connected series by means of perforations in their walls through which pass strands of cytoplasm. These perforations occur in groups known as *sieve-plates* in the end or lateral walls.

SILICULA. See *Siliqua*.

SILIQUA. Special type of capsule (q.v.) found in cabbages and related plants (family Cruciferae). Dry, elongated fruit formed from an ovary of two united carpels and divided by a central false septum between the carpels into two compartments (locules). Dehiscing by separation of carpel walls from below upwards leaving septum bounded by persistent placentas (*replum*), with seeds adhering to it. The Silicula has a similar structure but is short and broad, e.g. honesty, shepherd's purse.

SILURIAN. Geological period (q.v.), lasted approximately from 440 till 400 million years ago.

SINANTHROPUS. Peking man. Lower Pleistocene. Closely similar to (now put in same genus as) Pithecanthropus (q.v.).

SINUS. A space. (Bot.). Recess between two lobes of a leaf or other expanded organ. (Zool.). *Blood-sinus*, enormously expanded vein, found particularly in elasmobranchs. Applied also to haemocoel cavity of certain invertebrates. *Nasal sinus*, in mammals, air-filled space within certain bones of face, lined by mucous membraue, communicating with nasal cavity.

SINUSOIDS. Small blood-vessels which take the place of capillaries in some organs, notably liver and bone-marrow. Differ from capillaries in very various diameters (often much wider) and absence of regular endothelium. Macrophages (q.v.) lie in walls.

SINUS VENOSUS. Chamber of vertebrate heart, lying between veins and auricle(s); thin-walled; absent in adult birds and mammals. See *Pacemaker*.

SIPHONOPHORA. An order of coelenterates (class Hydrozoa) including the Portuguese man-of-war. Floating, colonial animals with extreme degree of specialization of members for, e.g. feeding, protection.

SIPHONOPODA. Cephalopoda (q.v.).

SIPHONOSTELE (SOLENOSTELE). Stele (q.v.) in which xylem and phloem form concentric cylinders around a central core of pith; *ectophloic*, with xylem cylinder surrounding pith and phloem surrounding xylem; *amphiphloic*, with phloem internal as well as external to xylem.

SIRENIA. Manatees, dugongs, seacows. An order of placental mammals. Aquatic (coastal) with transversely expanded tail, front legs as flippers, hind legs vestigial; vegetarian. Not closely related to whales (Cetacea) in spite of many similarities.

SKELETAL MUSCLE. Striped muscle (q.v.).

SLIDE. (1) Oblong of glass, usually 3 in. by 1 in. and 1–2 mm. thick, on which whole mounts, sections, etc., are placed for microscopical inspection. (2) The completed preparation made with such a piece of glass.

SLIME-FUNGI (SLIME-MOULDS). Myxomycetes (q.v.).

SMOOTH (PLAIN, INVOLUNTARY) MUSCLE. Contractile tissue of vertebrates, consisting of numbers of individual elongated spindle-shaped cells, with no transverse striations (see *Striped Muscle*), bound together by connective tissue fibres. Found mainly in sheets surrounding hollow organs, e.g. blood-vessels, gut. Controlled by autonomic nervous system. Undergoes contraction rather slowly, and has marked power of maintaining certain degrees of contraction (tonus). Slow-contracting muscle of many invertebrates, e.g. molluscs, is rather similar.

SOCIETY. (Bot.). Minor community within a consociation (q.v.) arising as a result of local variation in conditions and dominated by species other than consociation dominant.

SOIL PROFILE. Series of recognizably distinct layers (horizons) visible in vertical section through soil down to parent material. Study of soil profiles yields valuable information on character of soils.

SOLENOSTELE. Siphonostele (q.v.).

SOMATIC CELLS (SOMA). The cells of an organism, other than the germ-cells; see *Germ-Plasm*. *Somatic tissues*. Tissues other than those of viscera or blood-vessels; or tissues which surround body cavity (q.v.). *S. motor nerves* supply striped muscle; *s. sensory nerves* supply receptors of somatic tissues.

SOMITIC MESODERM. Mesoderm on dorsal side of vertebrate embryo, flanking the notochord in two longitudinal strips, each o which as development proceeds becomes segmented (q.v.) into a series of blocks, the *somites*. (In front of the ears however there is usually no segmentation into somites.) From each somite is derived a myotome (q.v.) innervated by one ventral nerve root (so that nerves are also segmentally arranged), and mesenchyme which forms connective tissue, and the vertebral column (latter also segmented). See *Lateral Plate; Pronephros.*

SOREDIA. Organs of vegetative reproduction in lichens (q.v.).

SORUS. (Bot.). Group, e.g. of fern sporangia.

SPADIX. Kind of inflorescence (q.v.).

SPATHE. Bract enclosing inflorescence of some monocotyledons, e.g. cuckoo pint.

SPECIALIZED. Having special adaptations to a particular habitat or mode of life which result in a wide divergence from presumed ancestral forms; and which tend to restrict the range of habitat which can be occupied and the variety of mode of life which can be followed and hence, it is assumed, to limit evolutionary flexibility. *Specialization*. (1) Acquisition of such special adaptations in the course of evolution. (2) Such a special adaptation.

SPECIATION. Origin of species.

SPECIES. The smallest unit of classification commonly used; i.e. the group whose members have the greatest mutual resemblance. The common names of familiar animals and plants often denote species e.g. man, fox, beech. For the great majority of animals and many plants, a species is roughly a group of individuals able to breed among themselves (if one disregards geographical separation) but not to breed with organisms of other groups. Consequently the members of a species form a reproductively isolated group, whose genes do not combine with those of outsiders, but are able to re-combine continually by sexual reproduction within the group. As a result no striking differences in genetic composition and in the characters controlled by genes occur within the species, though local differences, which are recognized in classification as a *sub-species*, may arise through reproductive isolation which is only partial or has recently occurred. Many of these sub-species are no doubt on the road to becoming full species, but are still capable of interbreeding within their species. Reproductive isolation however admits of degrees and so is not easy to apply rigorously to the problem. Furthermore, it is usually unknown whether two kinds of organism are in fact able to breed together in natural conditions. Consequently species are in practice usually determined by experts in the particular group, on the basis of degree of difference, which is often, though not always, a close reflexion of reproductive isolation. Reproductive isolation, however, cannot be applied at all as a criterion to many organisms. Unless, for instance, sexual reproduction with cross-fertilization freely occurs, the conception of species as a breeding unit breaks down; it does so therefore for bacteria and some plants. Species must in such cases always be determined on the basis of degree of difference, according to convenience; and while numerous species of such organisms are well-marked and undisputed, there can exist no general rule as to the minimum degree of difference separating any two species. A comprehensive definition of species applying to all kinds of organisms is in fact hardly possible. For naming of species see *Binomial Nomenclature*.

SPECIFIC. (1) Peculiar (to). (2) Characteristic of a species.

SPERM. Spermatozoon (q.v.).

SPERMATHECA (RECEPTACULUM SEMINIS). Organ in a female or hermaphrodite animal which receives and stores sperms from the mate.

SPERMATID. Animal cell resulting from the (second) meiotic division of a secondary spermatocyte (q.v.). At first a cell of normal shape, it undergoes extensive cytoplasmic changes and condensation of nucleus which convert it into a spermatozoon.

SPERMATIUM. Non-motile male sex-cell; present in red Algae, and in Fungi, in some members of the Ascomycetes and Basidiomycetes.

SPERMATOCYTE. (1) (Bot.). Cell which becomes converted into

spermatozooid (without intervention of cell division). Cf. *Spermatid.* (2) (Zool.). Cell which undergoes meiosis, and thereby forms spermatids (q.v.). *Primary spermatocyte* undergoes first of the two meiotic divisions. As a result it gives rise to two *secondary spermatocytes.* The latter undergo the second meiotic division, each forming two spermatids (q.v.). One primary spermatocyte thus produces four spermatozoa (one primary oocyte producing only one ovum). Cf. *Oocyte*, *Maturation of Germ-Cells.*

SPERMATOGENESIS. Formation of sperm. See *Maturation of Germ-Cells*, *Spermatocyte.*

SPERMATOGONIUM. Cell of animal gonad which undergoes repeated mitosis and eventually gives origin to spermatocytes (q.v.).

SPERMATOPHORE. Small packet of sperm, produced by some species of animals having internal fertilization, e.g. newt, cephalopod.

SPERMATOPHYTA (SPERMAPHYTA). Seed plants. Division of plant kingdom providing dominant flora of present day, including most trees, shrubs, herbaceous plants, grasses, etc. Possessing a highly organized plant body of stem, leaf and root, with a well developed vascular system. Heterosporous plants with dominant sporophyte and much reduced gametophyte generations. The sporophyte generation is the plant itself. The male gametophyte consists of the germinated pollen grain with its pollen tube by which the non-motile gametes are conveyed to the egg. The female gametophyte is retained on the sporophyte, nourished and protected within the *ovule*, which after fertilization of the egg within it becomes a *seed.* Within the seed, protected by seed coat, is a young sporophyte plant, the *embryo.*
Including two groups, *Gymnospermae* (q.v.), with ovules lying unprotected on the megasporophylls, and *Angiospermae* (q.v.), in which the megasporophylls, known here as *carpels*, enclose the ovules in the *ovary.* After fertilization, ovary becomes a fruit containing one or more seeds. In gymnosperms, sporophylls are usually arranged in cones; in angiosperms, sporophylls (stamens and carpels) are grouped in flowers.

SPERMATOZOID (ANTHEROZOID). (Bot.). Small motile male gamete with flagella.

SPERMATOZOON. (Zool.). Small, motile, usually flagellated male gamete (not flagellated in Nematoda, decapod Crustacea, diplopod Myriapoda, mites).

SPERMOGONIUM (SPERMAGONIUM, SPERMAGONE, PYCNIUM). (Of Fungi), flask-shaped or flattened, hollow structure in which spermatia are formed.

SPHENODON. The tuatara, a lizard-like New Zealand primitive reptile, sole living member of order Rhyncocephalia. Has a very well developed pineal eye (q.v.).

SPHENOPHYLLALES. Order of extinct Pteridophyta (q.v.) related to Equisetales (q.v.) which attained its maximum development in the Carboniferous and Permian periods (q.v.). Herbaceous or shrubby in habit, with grooved stem bearing whorls of leaves at nodes and terminating in a cone consisting of whorls of bracts bearing sporangiophores in their axils. Homosporous.

SPHINCTER. Ring of smooth muscle in wall of tubular organ or of an opening of a hollow organ, able by contraction to narrow or close lumen, e.g. pyloric sphincter (see *Pylorus*); anal sphincter.

SPIKE. Kind of inflorescence (q.v.).

SPINAL COLUMN. Vertebral column (q.v.).

SPINAL CORD. That part of the vertebrate central nervous system within the backbone. Consists of a hollow tube, but the hole is very small and the walls relatively and unevenly thick. Contains numerous nerve-cells and bundles of nerve fibres, particularly those connecting all levels of the spinal cord with the brain. Pairs of peripheral nerves, one nerve on each side per segment, leave the spinal cord to be distributed to the body. See *Nerve Root.* A great deal of simple co-ordination of movements of limbs, viscera, etc., is done in the spinal cord by reflexes (q.v.).

SPINAL NERVE. Peripheral nerve arising from spinal cord (q.v.) of vertebrate. One on each side in each segment. See *Nerve Root, Cranial Nerve.*

SPINDLE. Body formed within a cell at mitosis or meiosis, taking part in distribution of chromatids to the two daughter-nuclei. Commonly but not always ellipsoid. Appears at metaphase, with chromosomes arranged at its equator. See *Spindle Attachment.* Spindle is relatively solid (a gel) probably composed of longitudinally orientated protein molecules. Movement of chromatids apart during anaphase may be due to pulling by contractile fibres in spindle.

SPINDLE-ATTACHMENT (ATTACHMENT CONSTRICTION, CENTROMERE). Region of chromosome which attaches it to spindle (q.v.) at mitosis or meiosis. Unlike rest of chromosome, it is single during prophase of mitosis, and doubles only at metaphase. May be partly responsible for organizing the spindle. See Figs. 6A, 6B, pp. 147, 148.

SPINNERETS. Of spiders, four to six elevations at end of abdomen on to surface of which ducts of spinning glands open. Probably represent abdominal appendages.

SPIRACLE. (1) Dorsally situated remnant of first (most anterior) gill-slit in many fish. Its small size results from the connexion formed between mandibular and hyoid arches for firm attachment of jaws; between these arches the spiracle lies. See *Visceral Arch, Hyostylic Jaw Suspension.* Placoderms had full-sized first gill-slit. In most living bony fish, spiracle is closed up. The gill-pouch of embryo tetrapods representing spiracle develops into cavity of middle ear (q.v.) and Eustachian tube. (2) Of insects, synonym of stigma; external opening of trachea.

SPIRAL CLEAVAGE. A type of cleavage (q.v.). First two divisions are longitudinal. Third is latitudinal and unequal, and forms four small cells at animal pole which by an obliquity of their spindles lie above the grooves between the large vegetal cells. These four small cells are usually extruded in a clockwise direction in relation to the large cells. At the next division, the large cells again form four large and four small cells, and the original small quartet divides; again the spindles lie obliquely, but this time the direction in all divisions is anticlockwise. Clockwise and anticlockwise alternate, and a highly characteristic and constant pattern of blastomeres is formed. This peculiar method of cleavage must be an indication of evolutionary relationship between groups possessing it: polyclad Turbellaria, Annelida, Mollusca (except cephalopods), perhaps rotifers.

SPIRAL THICKENING. Internal thickening of wall of a xylem vessel or tracheid in the form of a spiral band. Occurs in cells of protoxylem and, whilst providing mechanical support, permits longitudinal stretching as neighbouring cells grow.

SPIRAL VALVE. Spiral fold of mucous membrane projecting into intestine of certain fish, e.g. elasmobranchs, ganoids, and dipnoi probably increasing absorptive surface.

SPIRILLUM. Corkscrew-shaped bacterium.

SPIROCHAETE. Elongated, spirally twisted organism, moving by flexions of the body, not by flagella. Usually classified with Bacteria. Some free-living, some parasitic and cause disease, e.g. syphilis.

SPLANCHNOCRANIUM. Part of skull comprising jaws and their attachments. Cf. *Neurocranium.*

SPLEEN. Mass of lymphoid tissue in mesentery of stomach or intestine of gnathostomes; but unlike lymph nodes is interposed in blood, not lymph, circulation. An important source of lymphocytes, and an important part of reticulo-endothelial system (q.v.) defending the bloodstream against invading organisms, and removing red blood cells at the end of their life. Acts also as a store of red blood cells; in emergency contraction of smooth muscle in spleen squeezes them into blood stream.

SPONGE. See *Porifera. Bath sponge* is the cleaned skeleton of certain sponges, consisting of interlocking fibres of a silk-like protein material, *spongin.*

SPONTANEOUS GENERATION. Present-day origin of organisms from non-living things, belief in which is now discredited. Until the rise of bacteriology, initiated by Pasteur, spontaneous generation was widely believed to be true for micro-organisms, which were supposed normally to develop from non-living organic material. See *Biogenesis.*

SPORANGIOPHORE. (Of Fungi), hypha bearing one or more sporangia; sometimes morphologically distinct from vegetative hyphae.

SPORANGIUM. (Bot.). Organ within which are produced asexual spores.

SPORE. Single-celled or several-celled reproductive body that becomes detached from the parent and gives rise either directly or indirectly to a new individual. Usually microscopic, of many different types and produced in a variety of ways. Thin or thick walled; often serving for very rapid increase in the population of the species, produced in enormous numbers and distributed far and wide by wind, water, animals; others are resting spores, the means of survival through an unfavourable period. Occurring in all groups of plants, particularly in Fungi; in Bacteria; and in Protozoa.

SPORE MOTHER CELL. (Bot.). Diploid cell giving rise by meiosis to four haploid spores.

SPOROGONIUM. Spore-producing structure of liverworts and mosses that develops after sexual reproduction; sporophyte generation of these plants.

SPOROPHORE. (Of Fungi), general term for a structure producing and bearing spores, e.g. sporangiophore, conidiphore (simple sporophores), mushroom (complex sporophore).

SPOROPHYLL. Leaf that bears sporangia. In some plants indistinguishable from ordinary leaves except by presence of sporangia, e.g. bracken fern; in others, much modified and superficially quite unlike ordinary leaves, e.g. stamens and carpels of flowering plants.

SPOROPHYTE. Phase of life-cycle of plants which has diploid (q.v.) nuclei, and during which spores are produced. It arises by union of sex-cells produced by (haploid) gametophytes (q.v.) See *Alternation of Generations*.

SPOROZOA. Class of Protozoa, all parasitic. Includes malaria parasite and many others of great economic and medical importance.

SPORT. Somatic mutation (q.v.), e.g. bud sport.

SPRINGTAIL. See *Collembola*.

SQUAMATA. Lizards and snakes. An order of Reptilia. Contains suborders Lacertilia (lizards) and Ophidia (snakes). Body covered with horny epidermal scales.

SQUAMOSAL. Membrane bone of skull, which in mammals takes over from quadrate the articulation with lower jaw. See *Ear Ossicle*.

STAMEN. Organ of flower which produces microspores (shed after development as pollen grains); a microsporophyll; comprising a stalk or *filament* bearing at its apex an *anther*. The anther consists of two lobes united by a prolongation of the filament (*connective*) and in each lobe there are two *pollen sacs*, containing pollen. See *Flower*.

STAMINATE. (Of flowers), possessing stamens but not carpels, i.e. male. Cf. *Pistillate*.

STAMINODE. Sterile stamen, one that does not produce pollen.

STAPES. One of the three mammalian ear ossicles (q.v.), stirrup-shaped because pierced by artery, representing columella auris of other tetrapods, hyomandibula of fish.

STARCH SHEATH. Innermost layer of cells of cortex in young angiosperm stems, containing abundant and large starch grains; considered homologous with endodermis and may sometimes lose its starch and become thickened as an endodermis at a later stage.

STATOBLAST. Resistant reproductive body, vegetatively produced, which is capable of withstanding adverse conditions; formed by fresh-water Polyzoa.

STATOCYST. Organ of balance consisting of a vesicle, containing granules of lime, sand, etc. (*statoliths*) which stimulate sensory cells as the animal moves; present in some Crustacea, flatworms, etc. Sometimes known as otocyst and otoliths (q.v.).

STATOCYTE. Plant cell that contains statolith(s).

STATOLITH. (1) Solid inclusion of a plant cell, commonly a starch grain, free to move under influence of gravity; thought by some to be the means whereby position with respect to gravity influences a plant organ. (2) See *Statocyst.*

STEGOCEPHALIA. Name (obsolete in classification) applied to fossil Amphibia; have complete bony covering to head in contrast to modern Amphibia.

STELE (VASCULAR CYLINDER). Cylinder or core of vascular tissue in centre of roots and stems, consisting of xylem, phloem, pericycle, and, in some steles, pith and medullary rays; surrounded by endodermis. Structure of the stele differs in different groups of plants. See *Protostele*, *Siphonostele*, *Dictyostele*, *Vascular Bundle.*

STEM. Normally aerial part of axis of vascular plants, bearing leaves and buds at definite positions (nodes) and reproductive structures, e.g. flowers. Some stems are subterranean, e.g. rhizomes but these, like all stems, are distinguished externally from roots by the occurrence of leaves (scale-leaves on rhizomes) with buds in their axils, and internally by having vascular bundles arranged in a ring forming a hollow cylinder, or scattered throughout tissue of the stem, and with protoxylem most commonly endarch.

STENOHALINE. Unable to tolerate wide variation of osmotic pressure of environment. Cf. *Euryhaline.*

STENOTHERMOUS. Unable to tolerate wide variation of temperature of environment. Cf. *Eurythermous.*

STERIGMA. (Of Fungi), minute stalk-bearing spore or chain of spores.

STERILE. (1) Unable to reproduce sexually. (2) Free from living micro-organisms.

STERNUM. (1) Breast bone. Bone of terrestrial vertebrates (tetrapods) in the middle of the ventral side of the chest, to which the ventral ends of most of the ribs are attached. At its anterior end attached to shoulder girdle (q.v.). (2) Of insects, cuticle on ventral side of each segment, often forming a thickened plate.

STEROIDS. Chemically similar but biologically diverse substances, e.g. bile acids; vitamin D; gonad and adrenal cortical hormones; some carcinogens; active parts of toad poisons and digitalis.

Solubility similar to fats. Saturated hydrocarbons containing seventeen carbon atoms in a system of rings, three six-membered and one five-membered, condensed together (six atoms being shared between rings); usually excluding however the true sterols (q.v.).

STEROLS. Compounds with the general chemical ring-structure of a steroid, but with certain specific features of structure, including a long side-chain and an alcohol group. Sterols occur universally in plants and animals. E.g. cholesterol, ergosterol.

STIGMA. (Bot.). Terminal expansion of style, surface of the carpel (q.v.) which receives pollen. (Zool.). Spiracle (q.v.) of insect.

STIMULUS. Any change in the environment of an organism or of part of it, which is intense enough to produce a change in the activities of the living material, without itself providing energy for the new activities.

STIPE. Stalk. (1) Of fruit-bodies of certain higher Fungi, e.g. mushroom. (2) Of thallus in sea weeds, e.g. bladder-wrack.

STIPULE. Small, usually leaf-like appendage, found one on either side of leaf-stalk in many plants, protecting axillary bud and often taking part in photosynthesis.

STOCK. Part of plant, usually consisting of the root system, together with a larger or smaller part of the stem, on to which is grafted a part of another plant (scion).

STOLON. (Bot.). Horizontally growing stem that roots at nodes, e.g. strawberry runner. (Zool.). Root-like part of colony of animals, e.g. hydroids, which fixes it to the substrate; in certain urochordates, outgrowth of body concerned with budding.

STOMA (PL. STOMATA). (1) Pore in the epidermis of plants, present in large numbers, particularly in leaves, through which gaseous exchange takes place. Each stoma is surrounded by two specialized, crescent-shaped epidermal cells known as *guard-cells*, whose movements, due to changes in turgidity, govern opening and closing of pore. (2) Includes both pore and guard-cells.

STOMACH. Enlargement of the anterior region of the gut. In vertebrates follows oesophagus, and usually has muscular walls which churn food, and lining cells which secrete pepsin and hydrochloric acid. Initial steps in digestion occur there, food held by pylorus (q.v.) until reduced to a mush.

STOMIUM. Place in wall of fern sporangium where rupture occurs at maturity, releasing the spores.

STOMODOEUM. An intucking of ectoderm meeting endoderm of anterior part of gut, forming mouth.

STONE-CELL (SCLEREID). Element of sclerenchyma (q.v.).

STRATUM CORNEUM. Outer layer of epidermis of land-living vertebrates, cells of which have been converted to keratin (q.v.). Responsible for proofing the body against entry of bacteria and ultra-violet rays, and against loss of water.

STRIATED MUSCLE. Striped muscle (q.v.).

STRIDULATION. Production of sounds by rubbing together of certain modified surfaces; e.g. of part of hind leg against part of forewing by grasshoppers. In some insects important in bringing sexes together.

STRIPED (STRIATED, SKELETAL, VOLUNTARY) MUSCLE. Contractile tissue consisting (in vertebrates) of large elongated cells (muscle-fibres) with many nuclei, the cytoplasm of which bears conspicuous striations at right-angles to long axis. Cytoplasm contains numerous longitudinal fibrils, each having alternating bands of different composition; the cross-striations of the whole muscle fibre are the result of similar bands of the fibrils lying side by side. A muscle fibre, on stimulation by its nerve, or (artificially) by direct action of electrical or mechanical stimuli, contracts by shortening and thickening. See *Myosin*. Muscle fibres are bound together into muscular tissue by connective tissue fibres. Muscles made of striped fibres undergo very rapid contraction and are particularly concerned with locomotion, by moving skeletal parts to which they are usually attached, in vertebrates, insects and members of some other groups. Cf. *Cardiac Muscle*, *Smooth Muscle*.

STROBILA. The 'body' of a tapeworm, consisting of a string of proglottides.

STROBILUS. Cone (q.v.).

STROMA. (1) Tissue-like mass of fungal hyphae in or from which fruit-bodies are produced. (2) Intercellular material or connective tissue component of an animal organ.

STRUCTURELESS LAMELLA. Mesogloea (q.v.).

STYLE. Prolongation of carpel supporting stigma. See *Carpel*.

SUBCUTANEOUS. Immediately below dermis of vertebrate skin. *S. tissue* is usually loose connective tissue, which may contain much fat (see *Blubber*, *Fascia*) and in many tetrapods a sheet of striped muscle (*panniculus carnosus*, conspicuous in some mammals) which can move the skin or its scales.

SUBERIN. Complex mixture of oxidation and condensation products of fatty acids; present in walls of cork cells rendering them impervious to water.

SUBERIZATION. Deposition of suberin (q.v.).

SUB-SPECIES. Subdivision of a species (q.v.) forming a group whose members resemble each other in certain characteristics, and differ from other members of the species, though there may be no sharp dividing line. Polymorphism (q.v.) is excluded. While breeding is possible and in many cases occurs between members of different sub-species of the same species, it does not occur as freely as within the confines of the sub-species. Because reproductive isolation is incomplete, sub-species nearly always grade into each other. The partial reproductive isolation is commonly due to the occupation of different geographical areas. Some sub-species are probably new species in the making. See *Binomial Classification.*

SUBSTRATE. (1) See *Enzyme.* (2) Ground or other solid object on which animals walk or to which they are attached.

SUCCESSION. (Of plants), progressive change in composition of plant population during development of vegetation, from initial colonization to attainment of climax.

SUCCULENT. Type of xerophytic plant that stores up water within its tissues; having a fleshy appearance, e.g. cacti, houseleek.

SUCCUS ENTERICUS. Digestive juice secreted by walls of small intestine of vertebrates. Contains numerous enzymes, e.g. erepsin, sucrase, lactase, and enterokinase.

SUCRASE. See *Sucrose.*

SUCROSE. Cane sugar. A disaccharide (with twelve carbon atoms) widespread in plants, but not in animals. Not found in mammalian body, except in food in gut. It is a compound of glucose and fructose (into which it is split by enzyme *sucrase*, also called *invertase* or *saccharase*).

SUMMATION. Generally, additive effect of separate stimuli. In neurophysiology, additive effect of separate impulses arriving at a nerve-cell or effector-cell. As a very simplified example, suppose that nerve-cell A has synapses with the nerve-fibres of nerve-cells B and C. Impulses arriving in B alone or in C alone fail to stimulate the production of an impulse in A. But if impulses arrive, simultaneously or very shortly after each other in both B and C, their joint action produces an impulse in A. The separation of the summating impulses may be *spatial*, that is, the impulses arrive at different synapses on the same nerve-cell; or *temporal*, that is, they arrive successively at the same cell; or both. Many impulses may be involved in any one process of summation. Temporal summation at a single synapse probably does not occur in vertebrates owing to the short duration of effect of each impulse which fades before the refractory period (see *Impulse*) will allow the next impulse to arrive. Besides summation which produces impulses (i.e. is excitatory), summation of inhibition (q.v.) also occurs. In general, summation is one of the main ways in which impulses can interact, so that nerve-cells in co-operation can produce effects quite different from the sum of those due to the nerve-cells singly. The result is the exquisite gradation of behaviour corresponding to the immense variety of the stimulus situations to which an animal reacts.

SUPERIOR OVARY. See *Receptacle.*

SUPINATION. See *Pronation.*

SUPRARENAL GLAND. See *Adrenal Gland.*

SUPRAVITAL STAINING. See *Vital Staining.*

SURVIVAL VALUE. The nature or degree of the effectiveness of a given characteristic in promoting the organism's ability to contribute offspring to the future population. See *Natural Selection.*

SUSPENSOR. (1) In phycomycete Fungi of the order Mucorales, cell supporting a gametangium. (2) In seed plants and in a few pteridophytes, structure developing from fertilized egg-cell, along with embryo, and bearing latter at its apex. By elongation of suspensor embryo is carried into nutritive endosperm or prothallus tissue.

SUTURE. Line of junction. (1) In flowering plants line of fusion of edges of a carpel is known as ventral suture. Mid-rib of carpel is known as dorsal suture, not implying any fusion of parts to form it but to distinguish it from the ventral (true) suture. (2) Junction between the irregular interlocking edges of certain contiguous skull bones in vertebrates; occupied by fibrous tissue. In man sutures slowly become obliterated during life by fusion of adjacent bones, allowing age at death to be roughly judged. (3) In insects, junctions between plates of hardened cuticle of exoskeleton. (4) (Surgical), to sew a wound together.

SUTURE-LINE. On shell of ammonite (q.v.), line marking junction of edge of septum with side wall of shell; often very complex; its changes of form during evolution are known in great detail.

SWARM-SPORE. Zoospore (q.v.).

SWEAT GLANDS. Skin glands of mammal secreting a dilute (hypotonic) solution of the salts and other small molecules present in blood. Function, cooling the animal by evaporation. Usually under nervous control of sympathetic system. Formed from epidermis, but project deep into dermis. Present in very variable numbers in different mammals, e.g. dog has very few, and those mainly on pads of feet.

SWIM-BLADDER (AIR-BLADDER). Bladder containing gas, present in roof of the abdominal cavity in bony fish (Actinopterygii). Functions as a hydrostatic organ, varying the specific gravity of the fish to match the depth at which it is swimming, so that the fish can remain there effortlessly. Derived from the lungs of ancestral fish, which were partly air-breathers. See *Lung.*

SYMBIONT. Symbiotic organism.

SYMBIOSIS. (1) Association of dissimilar organisms whatever the relationship between the two partners. (2) Association of dissimilar organisms to their mutual advantage, e.g. association of nitrogen-fixing bacteria with leguminous plants (peas, beans, etc.). The bacteria, inhabiting nodules on roots, manufacture nitrogen compounds from nitrogen of the air, which become available to the plant. From the latter bacteria obtain carbohydrates and other food materials.

SYMPATHETIC GANGLION. See *Autonomic Nervous System.*

SYMPATHETIC SYSTEM. (1) That part of autonomic nervous system (q.v.) sometimes called *orthosympathetic.* (2) Equivalent to autonomic nervous system.

SYMPATRIC. (Of geographical relationship of different species or sub-species) occurring together, i.e. with areas of distribution that coincide or overlap. Cf. *Allopatric.*

SYMPETALAE. Sub-class of Angiospermae in which the petals are united into a sympetalous (gamopetalous) corolla. Cf. *Archichlamydeae.*

SYMPETALOUS. Gamopetalous (q.v.).

SYMPHYSIS. Type of joint allowing only slight movement, in which the surfaces of the two articulating bones, both covered with a layer of smooth cartilage, are closely tied together by collagen fibres running between, e.g. pubic symphysis, joints between centra of vertebral column.

SYMPODIUM. Composite axis produced, and increasing in length, by successive development of lateral buds just behind the apex. Cf. *Monopodium.*

SYNANGIUM. Compound structure formed by lateral union of sporangia, present in certain ferns.

SYNAPSE. The nervous system of all animals contains immense numbers of distinct nerve-cells which touch each other only at certain places, synapses. There is no continuity between the nerve-cells at these places. Stimulation of one nerve-cell by another is probably confined to the synapses. A synapse is commonly formed by contact of the tip of a terminal branch of the axon belonging to one nerve-cell with the cell-body or with a dendrite of the other nerve-cell. Each axon (unless it is of a motor-fibre) usually has several synapses with each of several other nerve-cells; and each cell-body plus dendrites (unless it is a sensory-cell) has synapses with several nerve-fibres. The cell-body of a large vertebrate motor nerve-cell may bear several hundred synapses. An impulse, when it reaches a synapse, has to stimulate the next nerve-cell if it is to produce any further effect. This stimulation takes an appreciable time to occur (*synaptic delay*). Such stimulation may fail to occur if the impulse arrives from the wrong direction, since most synapses transmit only in one direction; thus sensory nerve-cells cannot usually be stimulated across synapses in the central nervous system. Furthermore the stimulation across a synapse may fail to produce an impulse in the nerve-cell receiving it, causing only a transitory rise in responsiveness. (See *Summation, Facilitation.*) Through the operation of summation synapses are highly flexible interconnexions. In those animals which have a C.N.S. most synapses are located there.

SYNAPSIS. See *Pairing.*

SYNCARPOUS. (Of the gynoecium of flowering plants), with united carpels, e.g. tulip. See *Flower.*

SYNCYTIUM. (Zool.). Mass of cytoplasm, enclosed in a single continuous plasma membrane, containing many nuclei. The cytoplasm may be in the form of a sheet (forming an epithelium, e.g. trophoblast of many mammalian placentas); of a cylinder, e.g.

striped muscle; of a network of almost discrete cells, each with a nucleus, but with cytoplasmic continuity through intercellular bridges, though it is difficult to establish this with certainty. The term is not used in botany. See *Cell, Plasmodium Coenocyte.*

SYNECOLOGY. Ecology of communities as opposed to individual species (*autecology*).

SYNERGIDAE. See *Embryo sac.*

SYNERGISM. (1) Combined activity of agencies, e.g. drugs, hormones, which separately influence a certain process in the same direction, such that an effect is produced greater than sum of effects of each agency acting alone. (2) Sometimes used for combined activity such that effect is either sum of the separate effects (summation), or greater than the sum of the separate effects (potentiation), it does not matter which; i.e. the agencies are not antagonistic to each other. Cf. *Antagonism.*

SYNGAMY. Union of gametes in fertilization.

SYNGENESIOUS. (Of stamens), united by their anthers, e.g. dandelion and other members of family Compositae.

SYNOVIAL MEMBRANE. Membrane of connective tissue lined with flattened cells (mesothelium) forming a bag (*Synovial sac*) enclosing a freely movable joint, e.g. elbow joint, being attached to the bones at either side of the joint. The bag is filled with a viscous fluid (*synovial fluid*) containing mucoprotein, lubricating the smooth cartilage surfaces which make the contact between the two bones.

SYRINX. Sound-producing organ of birds, situated at point where trachea splits into bronchi (in quite a different region from larynx, q.v.).

SYSTEMATICS. Study of the classification of living things, with emphasis on their evolutionary relationships.

SYSTEMIC. Generally distributed throughout an organism.

SYSTEMIC ARCH. Fourth aortic arch (q.v.) of tetrapod vertebrate embryo, becoming in adult main blood-supply for body other than head. In Amphibia and reptiles both right and left arches persist in the adult; in birds only the right; in mammals only the left (the aorta).

SYSTOLE. (1) Phase of heart-beat when heart muscle contracts, squeezing blood into arterial system. See *Pacemaker.* (2) Phase of contraction of contractile vacuole (q.v.). Cf. *Diastole.*

T

TACTILE. (Adj.). Of touch. *T. corpuscle.* Receptor end-organ of touch.

TAGMA (plural TAGMATA). Division, each of several segments,

into which body of arthropods is marked by differences in width, appendages, etc., e.g. head, throax, abdomen.

TANNINS. Group of astringent substances of wide occurrence in plants, dissolved in cell-sap; particularly common in the bark of trees, unripe fruits, leaves, and galls. Complex organic compounds containing phenols, hydroxy acids, or glucosides. Function in plant not known with certainty. Used in production of ink and leather.

TAPETUM. (1) In vascular plants, layer of cells, rich in food material, surrounding a group of spore mother cells, e.g. in fern sporangium; pollen sacs of anther; gradually disintegrate and liberate their contents which are absorbed by the developing spores. (2) Of vertebrate eye, reflecting layer of retina or choroid.

TAPEWORM. Parasitic flat-worm (Cestoda). Adult lives in gut of vertebrates; body long and ribbon-like, consisting of a chain of proglottides, attached to host by scolex. Eggs develop into six-hooked embryos (onchosperes) which pass out with faeces of host, and if eaten by suitable animal, develop into larval stage (cysticercus, cysticercoid, or plerocercoid, q.v.). This becomes sexually mature only when eaten by the definitive host, which may not occur until after it has parasitized another intermediate host.

TAP ROOT. Root system with a prominent main root, directed vertically downwards and bearing smaller lateral roots, e.g. dandelion; sometimes becoming very swollen, containing stored food material, e.g. carrot, parsnip. Cf. *Fibrous root.*

TARDIGRADA. Bear-animalcules. Minute animals found in moss, etc. Probably specialized arthropods.

TARSAL BONES (TARSALS). Bones of the proximal part of the hind-foot (roughly the ankle) in tetrapod vertebrates. (Cf. *Carpal bones* of fore-foot). Primitively 10–12 bones in a compact group: seven in man, one of them (calcaneum) forming the heel. Articulate on proximal side with tibia and fibula, on distal side with metatarsals. See Fig. 7, p. 175.

TARSIUS. A primitive primate, in some ways intermediate between lemurs and anthropoids, found in East Indies.

TARSUS. (1) Region of hind-leg of tetrapod vertebrates containing tarsal bones (q.v.); roughly the ankle. (2) One of the segments (the fifth from the base) of an insect leg.

TASTE-BUD. Receptor end-organ for taste. See *Papillae of Tongue.* No structural differentiation of buds corresponding to the four tastes (sweet, bitter, sour, salty) is known.

TAXIS. Locomotory movement of an organism or cell, e.g. gamete, in response to a directional stimulus, the direction of movement being orientated in relation to the stimulus, e.g. to gradient of temperature or to direction of illumination. Chemotaxis, geotaxis, phototaxis, according to nature of stimulus. Cf. *Tropism, Kinesis.*

TAXON. General term for a taxonomic group whatever its rank.

TAXONOMY. Science of the classification of living things. *Classical taxonomy* is concerned with description, naming and classification on the basis of morphology. *Experimental taxonomy* is concerned with analysing patterns of variation in order to discover how they evolved, with identification of evolutionary units, determining by experiment the genetical inter-relationships between them and the role of environment in their formation.

TELEOSTEI. A sub-class of Actinopterygii containing the great majority of existing fish (some 20,000 species).

TELEOSTOMI. A group of fish, now obsolete in classification, combining Actinopterygii and Crossopterygii, and excluding Dipnoi.

TELOPHASE. Terminal stage of mitosis (q.v.) or meiosis (q.v.) during which nuclei revert to resting-stage. See Figs. 6A, 6B, pp. 147, 148.

TELSON. The hindmost segment of the arthropod abdomen. In insects present only in the embryo.

TENDON. Cord or band of connective tissue attaching muscle, usually to a bone. Consists almost entirely of parallel collagen fibres.

TENDRIL. Stem, leaf, or part of leaf, modified as a slender, branched or unbranched, thread-like structure, used by many climbing plants for attachment to a support, either by twining round it, e.g. pea, grape; or sticking to it by means of an adhesive disc at the tip, e.g. virginia creeper.

TEPAL. Individual member of perianth (q.v.), particularly of perianths not clearly differentiated into corolla and calyx, e.g. tulip.

TERGUM. Thickened plate of cuticle on dorsal side of a segment of an arthropod.

TERMITE. White ant. See *Isoptera.*

TERPENE. Unsaturated hydrocarbon occurring in plant oils and resins.

TERRITORY. (Zool.). An area or volume of the habitat occupied by an individual or group, trespassers on which, if they belong to the same species, are attacked. A common feature of vertebrate behaviour; has been found in fish, reptiles, birds, mammals. Particularly (though not exclusively) concerned with breeding behaviour, a male often holding a territory alone at first, being subsequently joined by female(s) who then share(s) in its defence throughout the breeding period.

TERTIARY. Geological period (q.v.); lasted approximately from 70 till 1 million years ago.

TESTA. Seed coat. Protective covering of embryo of seed plants formed from integument(s); usually hard and dry.

TEST-CROSS. See *Double Recessive.*

TESTIS. Organ of animal which produces sperms. In vertebrates it also produces sex hormones. See *Androgen.*

TESTOSTERONE. An androgen (q.v.), probably the principal one produced within male vertebrates. A steroid.

TETRAD. (1) (Bot.). Group of four spores formed by meiosis within a spore mother-cell, e.g. formation of microspores within a microspore mother-cell. The process is often described as tetrad division. (2) Paired chromosomes of meiosis, after each chromosome has duplicated itself, and the pair is visibly four-stranded.

TETRADYNAMOUS. (Of stamens), six in number, four longer than the other two, e.g. wallflower.

TETRAPLOID. Having four times the haploid number of chromosomes in a nucleus. A form of polyploidy. See *Allotetraploid.*

TETRAPODA. Four-footed animals. A grouping of vertebrate classes sometimes used in classification. Includes Amphibia, reptiles, birds, and mammals, i.e. all the essentially land-living vertebrate classes. All characterized by two pairs of pentadactyl limbs (q.v.).

TETRASPORE. First cell of gametophyte generation in many brown and red Algae; formed in fours in a tetrasporangium following meiosis.

THALAMUS. (Bot.). Receptacle (q.v.) of flower. (Zool.). Part of the vertebrate fore-brain, a major sensory co-ordinating region.

THALLOPHYTA. Division of plant kingdom containing the most primitive forms of plant life. Characterized by a simple plant body (*thallus*) and varying from unicellular, microscopic forms, to multicellular forms such as large seaweeds, 60 to 70 metres in length. Although these latter show internal tissue differentiation, there is no differentiation of root, stem, and leaf as in higher plants. Asexual reproduction is by spores and sexual reproduction by fusion of gametes produced in sexual organs of various types but consisting essentially of single cells.

Containing two groups, *Algae*, characterized by presence of chlorophyll, and *Fungi*, by its absence. In addition there are *Lichenes*, a small group of composite organisms, each comprising an alga and a fungus, living together in symbiotic association. *Myxomycetes* or slime-moulds, and *Bacteria*, are often included in the Thallophyta, although the precise relationships of these two groups are obscure.

THALLUS. Simple vegetative plant-body of division Thallophyta, showing no differentiation into root, stem and leaf. Unicellular or multicellular, consisting of branched or unbranched filaments or more or less flattened, ribbon-shaped.

THERAPSIDA. Mammal-like reptiles. An extinct order of Reptilia. Lived from Permian to Triassic, the main reptile group until the appearance of the Dinosaurs. Important for being ancestral to mammals, the transition being well documented by fossils.

THERMONASTY. Response to a general, non-directional temperature stimulus, e.g. opening of crocus and tulip flowers with increase in temperature.

THEROPHYTES. Class of Raunkiaer's Life Forms (q.v.).

THIAMIN (ANEURIN, VITAMIN B_1). Vitamin of B group. Its phosphate is a co-enzyme (cocarboxylase), very widely (universally?) distributed in living organisms, concerned in carbohydrate metabolism. Synthesized by some bacteria, moulds, and flagellates, and by green plants, but required as a vitamin by many living things, including man and all vertebrates and insects tested. Deficiency in man causes beri-beri.

THIGMOTROPISM. Haptotropism (q.v.).

THORACIC DUCT. Main lymph vessel of mammals receiving lymph from trunk (including lacteals) and hind-limbs, running up the thorax close to the vertebral column, and discharging into vena cava superior or an associated vein.

THORAX. (1) In terrestrial vertebrates region of the body containing heart and lungs (chest). Only in mammals is it clearly marked off from the abdomen by the diaphragm. (2) In insects, the group of three segments behind the head which bears the three pairs of legs and (when present) the wings.

THREAD CELL (CNIDOBLAST). Very specialized cell found only in coelenterates (of the phylum Cnidaria, i.e. hydroids, jelly-fish, sea-anemones, corals; not in Ctenophora) and in a few animals which prey upon them and 'adopt' their thread cells. Several kinds exist, having different functions, e.g. stinging, adhesion. A thread cell forms within itself an extraordinarily complicated body (nematocyst) consisting of a bladder within which a long hollow thread lies coiled until on stimulation, e.g. by touch of prey, the thread is shot out, and attaches to prey, or in case of stinging type, injects poison into it. Stinging of jelly-fish is due to thread cells present in enormous number of their tentacles.

THREADWORMS. Nematoda (q.v.).

THRESHOLD. That intensity of stimulus below which there is no response by a given irritable tissue.

THROMBIN. See *Blood-Clotting*.

THROMBOCYTE. See *Blood Platelet*.

THYMONUCLEIC ACID. Nucleic acid (q.v.) of desoxyribose type. First extracted from thymus, which is rich in nuclei.

THYMUS. Organ of vertebrates, usually in pharyngeal or neck region, formed embryologically from gill-pouches or gill-clefts. Function unknown, though often suspected to be an endocrine gland. In mammals lies in chest (in mediastinum), and the cells from gill-pouches are mixed up with masses of lymphocytes; the organ reaches maximum size at puberty, and thereafter slowly atrophies.

THYROGLOBULIN. A protein containing thyroxin. The form in which thyroid secretion is stored in thyroid gland, but not certainly the hormone itself.

THYROID GLAND. An endocrine gland of vertebrates. Unpaired organ in neck region, formed embryonically from endodermal cells of floor of pharynx, homologous with endostyle of amphioxus and ammocoetes. Secretes an iodine-containing hormone (see *Thyroxin*, *Thyroglobulin*), principal effect of which is to increase rate of oxidative processes. In amphibian tadpoles secretion of thyroid initiates metamorphosis. Thyroid gland is itself controlled by pituitary (thyrotrophic hormone). A trace of iodine in the diet is necessary to supply the thyroid, and deficiency of iodine produces enlargement (hyperplasia) of thyroid (one kind of goitre in man). Lack of the hormone from whatever cause, in human infants, produces mental and physical stunting (*cretinism*).

THYROTROPHIC (THYROTROPIC) HORMONE. Hormone secreted by anterior lobe of pituitary, stimulating secretory activity and, in high doses, growth of thyroid gland.

THYROXIN. An iodine-containing amino-acid, probably the thyroid hormone.

THYSANOPTERA. Thrips. An order of minute exopterygote insects most of which feed on sap of plants. Some are serious pests.

THYSANURA. See *Apterygota*.

TIBIA. (1) Shin-bone. The anterior of the two long bones (other is fibula) of the shank (below the knee) of the hind limb of tetrapod vertebrates. See Fig. 7, p. 175. (2) One of the segments (fourth from base) of an insect leg.

TICKS. Members of two families of the order Acarina. Blood suckers, some important carriers of disease, e.g. cattle fever tick.

TILLER. (Of grasses), side shoot arising at ground level.

TISSUE. A region consisting mainly of cells of the same sort (performing the same function) associated in large numbers, bound together by cell-walls (plants) or by intercellular material (animals); e.g. cells of plant cortex with their cellulose walls; or the numerous striped muscle fibres, bound together by collagen, with associated blood and lymph vessels, nerves and connective tissue cells, which make up muscle tissue.

TISSUE CULTURE (EXPLANTATION). A technique for maintaining fragments of animal or plant tissue or separated cells alive after their removal from the organism. The tissue fragments (*explants*) or cells are kept usually within some sort of glass vessel, in a medium of the right properties (isotonic, buffered to correct pH, with correct balance of metallic ions, see *Physiological Saline*, with suitable mechanical properties, at correct temperature, and containing the necessary oxygen and food material). Foreign organisms, especially bacteria and fungi, must usually be excluded. The medium may have to be frequently renewed to remove excretions and maintain food supply. The technique cannot be used for large pieces of tissue, because the cells in the middle of a large mass are not sufficiently close to the medium to receive food and dispose of excretions. The specific physiological functions, or in the case of embryonic tissues their specific course of development, may continue in tissue cultures.

TOADSTOOL. Popular name for fruit-bodies of fungi other than mushrooms belonging to the family Agaricaceae of the Basidiomycetes; assumed generally, though often wrongly, to be poisonous.

TONOPLAST. (Bot.). Inner plasma-membrane, bordering vacuole.

TONSIL. Mass of lymphoid tissue in mouth or pharynx of tetrapods. Lies close underneath mucous membrane, deep crevices of which may communicate with interior of tonsil. Man has a pair of palatine tonsils (at junction of mouth and pharynx) and a single pharyngeal tonsil ('adenoids' at back of nose). Probably concerned in defence of mouth and pharynx against bacteria.

TONUS (TONE). Continuous but usually moderate, physiological activity of a tissue or organ; e.g. striped muscle, as a result of continuous nervous stimulation, is normally in a state of moderate contraction (tonus), by which the posture of the animal is maintained.

TORUS. (1) Receptacle q.v.) of flower; (2) Thickened portion of closing membrane of bordered pits (q.v.).

TRACE ELEMENT. An element which must be available to an organism for its normal health though it is necessary only in minute amounts. E.g. higher plants need traces of at least the elements zinc, boron, manganese, molybdenum, and copper; lack of these may produce economically serious disease, such as 'heart-rot' of sugar beet (boron deficiency). In animals such deficiency disease is also known, e.g. 'coast disease' of cattle and sheep in Australia from lack of cobalt. In man, thyroid deficiency (goitre, cretinism) may be due to lack of semi-trace element, iodine. See *Thyroid*. Trace elements are probably constituents of enzyme systems (cf. vitamins); and also, in animals, of hormones.

TRACER. The atoms of most chemical elements are not all alike, but are of a few different kinds, called isotopes, differing in atomic weight but not in chemical properties. Some isotopes are very rare naturally, and these can be concentrated; and many radioactive ones can be prepared artificially. These isotopes are used experimentally as 'tracers'; i.e. they can be incorporated into compounds of biological importance, administered to an organism, and their movements and changes of chemical combination determined by analysis of the organism, or of its products. The radioactive isotopes are particularly easy to follow by their radiation; for instance by autoradiographs (q.v.).

TRACHEA. (1) (Bot.). Conducting element of xylem. See *Vessel*. (2) (Zool.). A tube which is part of the breathing apparatus of air-breathing animals. In land-living vertebrates the 'wind-pipe', a single tube which leads from the throat, starting at the glottis through the neck to the point where it bifurcates into two bronchi. In insects, tracheae ramify throughout the body and conduct air from a few openings at the surface (spiracles) directly to the tissues; made of epidermis and cuticle; finest branches are called tracheoles.

TRACHEAL GILL. Of aquatic insect larvae; organ whose thin-walled surface in contact with water allows interchange of respiratory gases between it and the tracheae it contains.

TRACHEID. Non-living element of xylem (q.v.) formed from a single cell. Elongated, with tapering ends and with thick, lignified, and pitted walls. It is a long, empty, firm-walled tube running parallel with long axis of the organ in which it lies, overlapping and in communication with adjacent tracheids by means of pits. Functioning in water conduction and in mechanical support. Cf. *Vessel.*

TRACHEOLE. Terminal branch of trachea (q.v.) of insects, pervading tissues in very large numbers. Interchange of oxygen and carbon dioxide between tissues and gas in tracheoles takes place by diffusion across their walls.

TRACHEOPHYTA. Classification term introduced as cognate with Thallophyta and Bryophyta to include all vascular plants (Pteridophyta and Spermatophyta); emphasizes physiological and phylogenetic importance of the vascular system. Widely used in modern American texts.

TRACT. Spatially delimited bundle of nerve fibres, latter usually all with similar connexions, in the central nervous system. Tracts connect nuclei with each other and with peripheral nervous system.

TRANSDUCTION. Transfer of genetic material from one bacterium to another through the agency of bacteriophage (q.v.). Gene or genes of one (host) bacterial cell become incorporated in phage particles which after release from dead host cell act as vectors in the transport of this genetic material to other bacterial cells.

TRANSECT. Line or belt of vegetation selected for charting plants; designed to study changes in composition of vegetation across a particular area.

TRANSFORMATION. Phenomenon in which certain bacteria, when grown in the presence of killed cells, culture filtrates or extracts from other, related strains, acquire some of the genetic characters of these strains (i.e. they are transformed). Transforming principle is DNA.

TRANSFUSION TISSUE. Tissue of empty cells with pitted and, occasionally, internally thickened walls, and protein-containing, parenchymatous cells, accompanying vascular tissue in leaves of most gymnosperms, lying on either side of the vascular bundles of the single vein. Thought to represent an extension of the vascular system, taking the place of lateral veins.

TRANSLOCATION. (1) (Bot.). Movement of soluble organic food materials through tissues of higher plants, e.g. from leaves to actively growing parts and storage organs, and from storage organs to actively growing parts. (2) (Genetics). Transfer of part of a chromosome into a different part of a homologous, or into a non-homologous, chromosome. See *Mutation.*

TRANSPIRATION. Loss of water-vapour by land plants. Occurs mainly from leaves and differs from simple evaporation in that it takes place from living tissue and is therefore influenced by physiology of plant. Transpiration takes place chiefly through stomata (q.v.) and to a much less extent through the cuticle. Its significance is not completely understood. So long as stomata are open during interchange of gases between plant and atmosphere in photosynthesis and respiration, loss of water-vapour to the atmosphere must occur, i.e. for a healthy plant transpiration is inevitable. When it is excessive it is often harmful to the plant, causing wilting and even death; on the other hand there seems little doubt that it facilitates the upward movement of mineral salts in the xylem in the *transpiration stream*, and it may also be of importance in preventing overheating in direct sunlight.

TRANSPLANTATION. (Zool.). Artificial removal of part of an organism from its normal position to an abnormal position in the same or another organism. Practically synonymous with grafting (see *Graft*) except that no close union with tissues of new position is necessarily implied.

TRANSVERSE PROCESS. Lateral projection, one on each side, of the neural arch of vertebra of tetrapod, with which head of rib articulates.

TREMATODA. (Flukes.) Class of Platyhelminthes. Parasites, with cuticle, two suckers, and forked gut. Usually complicated life-cycle. See *Fluke.*

TRIASSIC. Geological period (q.v.), lasted approximately from 225 till 180 million years ago.

TRIBE. Minor group used in classifying plants. Used in large families for groups of closely related genera within the family. Name of tribe ends in *-eae.*

TRICHOGYNE. In some green and red Algae, Ascomycete Fungi, and lichens, uni- or multicellular projection from female sex organ that receives male gamete or male nuclei before fertilization.

TRICHOME. Hair (q.v.). In blue-green Algae (*Cyanophyceae*), filament consisting of uniseriate or multiseriate chain of cells.

TRICHOPTERA. Caddis flies. Order of endopterygote insects with hairy bodies and wings; aquatic larvae live in cases which they make of wood, sand, leaves, etc.

TRICUSPID VALVE. Valve between right auricle and right ventricle of mammalian heart, consisting of three membranous flaps.

TRIGEMINAL NERVE. Fifth cranial nerve of vertebrates. In mammals mainly sensory, innervating teeth and skin of face. A dorsal root.

TRILOBITA. Class of Arthropoda. Now extinct but abundant from Cambrian to Silurian. Marine, superficially resembling woodlice; one pair of antennae, all other appendages biramous and similar to each other. Probably related to ancestors of Crustacea.

TRIPLOBLASTIC. Having the body made up of three layers (ectoderm, mesoderm, endoderm). As in all Metazoa except coelenterates, which are diploblastic (q.v.). See *Germ-Layers.*

TRIPLOID. Having three times the haploid number of chromosomes in a nucleus. A form of polyploidy.

TRISOMIC. Diploid but with one extra chromosome, homologous

with one of the existing pairs; so that one kind of chromosome is represented three times.

TROCHANTER. (1) A prominence on the femur of vertebrates to which muscles are attached. There are three trochanters on each femur in mammals. The largest one in man is the conspicuous bony prominence at the hip joint. (2) One of the segments (second from base) of an insect leg.

TROCHLEAR NERVE. Fourth cranial nerve of vertebrates. Almost entirely motor, supplying eye-muscle (superior oblique). A ventral root.

TROCHOPHORE (TROCHOSPHERE). Larva of Polychaeta, Mollusca and Rotifera. Ciliated, usually planktonic. Develops by spiral cleavage. The main band of cilia encircles body in front of mouth. Coelom arises in blocks of mesoderm, i.e. not as pouches of gut. Two protonephridia sometimes present.

TROPHOBLAST. Embryonic epithelium which encloses all embryonic structures of placental mammal, forming outer layer of chorion (q.v.), and establishing close contact with maternal tissues. Forms embryonic side of placenta (q.v.); selectively permeable and hormone-secreting.

TROPISM. (1) Response to stimulus, e.g. gravity or light, in plants and sedentary animals by growth curvature, the direction of curvature being determined by the direction from which the stimulus originates. Cf. *Nastic Movement*. (2) (Zool.). Equivalent to taxis (q.v.) but this usage is becoming obsolete.

TRUFFLE. Subterranean fruit-body of Ascomycete Fungi belonging to order Tuberales; prized as a gastronomic delicacy.

TRYPANOSOMA. Genus of flagellate Protozoa, parasitic in the blood of vertebrates, and in the gut of tsetse flies and other insects, which transmit them to vertebrates. Trypanosomes cause important diseases in man (sleeping sickness) and horses and cattle in Africa.

TRYPSIN. Enzyme (peptidase, q.v.) splitting proteins and peptides at certain specific peptide links, in alkaline solution. Secreted by vertebrate pancreas. Trypsin-like enzymes occur in digestive juices of most animals, and also intracellularly (cathepsin, q.v.) in various tissues.

TRYPSINOGEN. Almost inactive form in which trypsin is secreted by pancreas; a compound of trypsin and a polypeptide. Converted to trypsin by enterokinase (q.v.) and by trypsin already formed.

TSETSE FLY. Genus (*Glossina*) of dipteran insects. Blood-sucking flies which are responsible for carrying trypanosome diseases, e.g. sleeping sickness.

TUBE FEET (PODIA). Of Echinodermata, hollow extensile appendages connected to the water vascular system (q.v.). In some groups (starfish, sea-urchins) they end in suckers and are locomotor, in others (brittle stars, crinoids) suckers are absent. Those around the mouth may be used for feeding (sea cucumbers, brittle stars). In crinoids they are ciliated and are part of food-collecting mechanism.

TUBER. Swollen end of underground stem bearing buds in axils of scale-like rudimentary leaves (stem tuber), e.g. potato, or swollen root (root tuber), e.g. dahlia. Tubers contain stored food material and are organs of vegetative propagation.

TUNICA CORPUS CONCEPT. See *Apical Meristem.*

TUNICATA. Urochordata (q.v.).

TURBELLARIA. Eddyworms, a class of Platyhelminthes. Mostly aquatic, free-living. Ciliated epidermis, gut a simple or branched sac.

TURGOR. State of cell in which the cell-wall is rigid, stretched by increase in volume of vacuole and protoplasm during absorption of water. The cell is described as turgid.

TURION. Detached winter bud by means of which many water plants survive winter.

TYLOSE (TYLOSIS). Balloon-like enlargement of the membrane of a pit in the wall between a xylem parenchyma cell and a vessel or tracheid, protruding into cavity of latter and blocking it. Tyloses occur in wood of various plants, often abundantly in heartwood of trees, and may be induced to form by wounding.

TYMPANIC CAVITY. See *Ear, Middle.*

TYMPANIC MEMBRANE. Ear-drum. Double layer of epidermis with connective tissue between. See *Ear, Middle* and *Outer.*

TYPE SPECIMEN (HOLOTYPE). Original specimen(s) from which a description of a new species is made. When the exact specimen is not known or is lost, one has to be selected anew, and is called the *lectotype* (q.v.) or *neotype* (q.v.). See *Binomial Nomenclature.*

U

ULNA. The posterior of the two bones (other is radius) of fore-arm of tetrapod vertebrate fore-limb. Articulates with side of hand (forefoot) opposite to that on which thumb is. See Fig. 7, p. 175.

ULTRACENTRIFUGE. High speed centrifuge capable of sedimenting particles as small as a protein molecule. Rate of sedimentation is used to estimate molecular weight or particle size. Since different kinds of protein sediment at different rates, provides a means for determining whether a solution of protein is a mixture or not.

UMBEL. Kind of inflorescence (q.v.).

UMBILICAL CORD. Stalk projecting from ventral surface of embryo of placental mammal, in most species connecting it to placenta. Consists mainly of mesoderm and blood-vessels of allantois, containing also part or all of the cavities of yolk-sac and allantois.

Cord is surrounded by amniotic cavity. Usually breaks or is broken through at birth. See Fig. 8, p. 184.

UNGULATE. Hoofed mammal; usually adapted for running on firm, open ground, herbivorous, living in herds. Since hooves have appeared independently in several groups of mammals, Ungulata is a term obsolete in classification. Most ungulates belong to orders Artiodactyla and Perissodactyla.

UNGULIGRADE. Walking on tips of digits, which have hooves, e.g. horse, cow. See *Artiodactyla*, *Perissodactyla*. Cf. *Digitigrade*, *Plantigrade*.

UNICELLULAR. Of organisms, consisting of one cell (q.v.) only, as distinct from multicellular; e.g. Protozoa.

UNIOVULAR TWINS. Monozygotic twins (q.v.).

UNISEXUAL. (1) (Of flowering plants or flowers) having stamens and carpels in separate flowers. Can be either monoecious or dioecious. Cf. *Hermaphrodite*. (2) (Of an individual animal) producing either male or female gametes, but not both. Cf. *Hermaphrodite*.

UREA. Main excreted product of protein break-down in ureotelic (q.v.) vertebrates. Occurs also in plants. A nitrogen-containing organic compound, $CO(NH_2)_2$, readily soluble in water.

UREASE. Enzyme which splits urea into ammonia and water. Present in many plants, but amongst animals only in a few invertebrates, e.g. Crustacea.

UREDINALES. Rust fungi. Order of Basidiomycetes comprising obligate parasites of higher plants with succession of several different spore forms in complicated life cycle that in some species involves two host species. Many rust fungi are of great importance as plant pathogens, most important being *Puccinia graminis*, cause of black stem rust of wheat and other cereals. See *Fungi*, *Autoecious*, *Heteroecious*, *Physiologic Specialization*.

UREOTELIC. Excreting urea as main break-down product of amino-acids; ammonia from the latter (see *Deamination*) being converted to urea by ornithine cycle (q.v.). Characteristic of those vertebrates during whose embryonic development excretory products can freely diffuse away in this soluble form, i.e. those with aquatic or near-aquatic development (fish, Amphibia, Chelonia) and fully viviparous forms (mammals). Aquatic invertebrates excrete ammonia. Cf. *Uricotelic*.

URETER. Duct conveying urine away from kidney. In vertebrates term is usually restricted to duct leading from kidney (metanephros, q.v.) of amniote to urinary bladder; formed as outgrowth of Wolffian duct (q.v.).

URETHRA. Duct leading from urinary bladder (q.v.) of mammals to exterior. Joined by vas deferens (q.v.) in male.

URIC ACID. Complex nitrogen-containing organic compound, only slightly soluble in water. Main excreted product of breakdown of amino-acids and nucleic acids in uricotelic (q.v.) animals. Excreted by primates and Dalmatian dogs but not by other mammals as product of break-down of nucleic acids. Cf. *Ureotelic.*

URICOTELIC. Excreting uric acid as main break-down product of amino-acids. Characteristic of terrestrial animals which develop within a shell, unable to get rid of their excretory products, which they store in this insoluble form (squamata, birds, insects, and terrestrial gastropods). See *Cleidoic egg.* Cf. *Ureotelic.*

URINARY BLADDER. Sac storing urine; an expansion of the kidney duct in, e.g. Crustacea, teleosts. In tetrapods, a ventral diverticulum of hind end of gut (cloaca) which in embryo of amniotes leads into allantois (q.v.). Absent in adult birds and most reptiles.

URINIFEROUS TUBULE. Narrow coiled tube of vertebrate kidney leading from Bowman's capsule (q.v.) to the collecting ducts which convey the urine to the ureter or to the Wolffian duct. The fluid filtered into the Bowman's capsule is altered in composition as it travels along the tubule, valuable substances in it such as glucose, much of the salts, and in land vertebrates much of the water, being absorbed by the tubule walls; while some other substances are excreted into it. Blood supply to tubule is from renal portal system in anamniotes; from renal artery via glomerulus (q.v.) in amniotes.

UROCHORDATA (TUNICATA). Sea-squirts, etc. A group of marine chordate animals. See *Protochordata.* Adult sedentary (some secondarily free swimming), feeding by ciliary currents; gill slits; reduced nervous system; no notochord. Larva active, tadpole-like, with well-developed nervous system and notochord.

URODELA. Newts and salamanders; an order of the class Amphibia. Distinguished from the other orders, Anura and Apoda, by elongated body, with tail and normal limbs.

UROSTYLE. Unsegmented caudal region of vertebral column of Anura.

UTERINE TUBE. Fallopian tube (q.v.).

UTERUS. Womb. Muscular expansion of Müllerian duct (q.v.) of female mammals (excluding monotremes) in which the embryo develops. Usually paired, one connected with each Fallopian tube; but may, as in man, be single through fusion of lower part of Müllerian ducts in the embryo. Connected by vagina to exterior. Has glandular lining membrane (endometrium, q.v.; decidua, q.v.) which nourishes early embryo; and smooth muscle in its wall, which greatly increases in amount during pregnancy and whose contraction ultimately expels embryos and their placentae. Under control of sex hormones, it varies much in structure, enlarging at sexual maturity or in breeding season, diminishing during sexual inactivity; and in many mammals fluctuating in structure with oestrous cycle. See *Cervix, Placenta.*

UTRICLE (UTRICULUS). See *Ear, Inner.*

V

VACUOLE. Fluid-filled space within the plasma-membrane of a cell. A single vacuole, taking up most of volume of cell, is present in many plant cells; contains solution (cell sap) isotonic with cytoplasm. Animal cells usually have only minute vacuoles. See *Contractile Vacuole.*

VAGINA. Duct of female mammal (excluding monotremes) connecting uteri with exterior via a short vestibule; receives penis of male during copulation. Embryologically derived either from terminal part of Müllerian duct or from cloaca. Usually single and median, owing to fusion of lower part of Müllerian ducts in the embryo, but paired in some marsupials. Lined with stratified non-glandular-epithelium. In many mammals undergoes cyclical changes with oestrous cycle (q.v.) and in mice recognition of these changes (from scrapings of vaginal wall known as *vaginal smear* technique) are of great importance in work on sex hormones. See *Oestrogen.*

VAGUS NERVE. Tenth cranial nerve of vertebrates. In mammals mainly parasympathetic (q.v.) supplying oesophagus, stomach, heart; and sensory, from, e.g. lungs and heart. A dorsal root.

VALVE. (Bot.). (1) One of several parts into which capsule (q.v.) separates after dehiscence by longitudinal splitting. (2) applied also to each half of wall of diatom cell (frustule, q.v.).

VARIATIONS. Differences between individuals of same species, apart from those referable to life cycle. May be due to differences in environment during development, and/or differences in the genes, or cytoplasmic equipment with which the individual starts life.

VARIEGATION. Irregular variation of colour of plant organs, such as leaves and flowers, due to suppression of normal pigment development, e.g. chlorophyll, anthocyanin, in certain areas. An inherited character or a disease symptom, e.g. mosaic diseases due to virus infection.

VARIETY. (Bot.). The taxon (q.v.) below subspecies; a group which distinctly differs, for various reasons, from other varieties within the same sub-species. Often used loosely in Botany and Zoology to mean a variation of any kind within the species.

VASCULAR. Containing, or concerning, vessels which conduct fluid. In animals, the fluid is usually blood; in plants water, mineral salts, and synthesized food materials.

VASCULAR BUNDLE. Longitudinal strand of conducting (vascular) tissue consisting essentially of xylem and phloem; unit of stelar structure in stems of gymnosperms and angiosperms and occurring in appendages of stem, e.g. veins in leaves. In stems of gymnosperms and dicotyledons, arranged in a ring surrounding the pith, in monocotyledons, scattered throughout tissue of stem. May be (*a*) *collateral*, with phloem on same radius as xylem and external to it;

typical condition in angiosperms and gymnosperms. (*b*) *Bicollateral*, with two phloem groups, external and internal to xylem, on same radius; uncommon, occurring in some dicotyledons, e.g. marrow. (*c*) *Concentric*, with one tissue surrounding the other. Concentric bundles are *amphicribral* when phloem surrounds xylem, e.g. in some ferns, and *amphivasal* when xylem surrounds phloem, e.g. in rhizomes of certain monocotyledons. Vascular bundles are further described as *open*, when cambium is present, e.g. in most dicotyledons, and *closed*, when cambium is absent, e.g. in monocotyledons.

VASCULAR CYLINDER. See *Stele*.

VASCULAR PLANT. Plant possessing vascular system (q.v.), member of Pteridophyta or Spermatophyta.

VASCULAR SYSTEM. (1) (Bot.). Plant tissue consisting mainly of xylem and phloem which forms a continuous system throughout all parts of higher plants. It functions in conduction of water, mineral salts, and synthesized food materials and for mechanical support. Older stems and roots of perennial plants consist largely of vascular tissue. (2) (Zool.). Blood system (q.v.) or Water Vascular System (q.v.).

VAS DEFERENS. Tube (one on each side) conveying sperm from testis to exterior. In male amniote vertebrate conducting sperm from epididymis (q.v.) to cloaca or urethra (q.v.). See *Wolffian Duct*, *Vas Efferens*.

VAS EFFERENS. Tube (of which there are many on each side) conveying sperm from testis of vertebrate to mesonephros or epididymis.

VASOCONSTRICTION. Narrowing of blood-vessels; usually refers to arterioles, by contraction of smooth muscle in their walls, mainly controlled by sympathetic nervous system. Capillaries constrict apparently by change in shape of endothelial cells.

VASODILATATION. As vasoconstriction but *expansion* of blood-vessels, usually by *relaxation* of muscle.

VASOMOTOR. Concerned with constriction and dilatation of blood-vessels. *V. nerves*, sympathetic nerves of vertebrates controlling and maintaining tonus of smooth muscle of arterioles. There are separate constrictor and dilator nerves.

VECTOR. An animal which transmits parasites, e.g. mosquitoes are vectors of malaria parasite.

VEGETAL POLE (VEGETATIVE POLE). (Zool.). Point on surface of egg furthest from the egg nucleus, and usually at yolkiest end of the egg. See *Animal Pole*.

VEGETATIVE REPRODUCTION (V. PROPAGATION). Asexual reproduction in plants by detachment of some part of the plant body and its subsequent development into a complete plant, e.g. gemmae, rhizomes, bulbs, corms, tubers. Vegetative reproduction or propagation in animals is synonymous with asexual reproduction (q.v.).

VEIN. (1) (Bot.). Vascular bundle (q.v.) of a leaf. (2) (Zool.). Blood-vessel (of fairly small diameter to distinguish it from a venous *sinus*, q.v.) carrying blood from the minute vessels which supply tissues (the capillaries) to the heart. In vertebrates, veins have a smooth lining of flat cells (endothelium, q.v.) with muscular and fibrous tissue outside; compared with arteries their walls are thin and their internal diameter large; unlike arteries and capillaries they contain simple valves, which permit blood-flow only towards the heart. (3) Of insect wings, fine tubes of toughened cuticle which make a supporting framework. Contain air-tubes (tracheae, q.v.) nerves and blood.

VELAMEN. Water-absorbing tissue occurring on the outside of aerial roots in certain plants, e.g. epiphytic orchids, consisting of several layers of dead cells often with spirally thickened and perforated walls, which act as a sponge, soaking up water that runs over it.

VELIGER. Larva of Mollusca, which develops from a trochophore. Ciliated bands have become larger; foot, mantle, shell, and other organs of adult are present.

VENA CAVA INFERIOR (POSTERIOR VENA CAVA, POSTCAVAL VEIN). Main vein of tetrapod vertebrates passing blood into the heart from veins of almost all body behind the fore-limbs. A single median vein, largest in the body. Has no homologue in fish, whose posterior cardinal veins (q.v.) serve same purpose.

VENA CAVA SUPERIOR (ANTERIOR VENA CAVA, PRECAVAL VEIN). Main vein of tetrapod vertebrates returning blood to the heart (to right auricle) from the fore-limbs and head. Usually a pair, but in many mammals, including man, only the right one persists in adult. Homologous with ductus Cuvieri of fishes.

VENATION. (1) Arrangement of veins in a leaf. (2) System of veins (q.v.) in wing of insect. An important characteristic for classification.

VENTER. (Bot.). Swollen basal region of an archegonium, containing egg-cell.

VENTRAL. Situated at, or relatively nearer to, that side of the animal which (if not in the animal, at least in the group to which the animal belongs) is normally directed downwards with reference to gravity. In human beings, ventral side is directed forwards. Opposite of dorsal. *V. aorta*. See *Aorta*, *Ventral*. (Bot.). Also used of leaves, synonymous with Adaxial (q.v.).

VENTRAL ROOT (ANTERIOR ROOT, MOTOR ROOT). Nerve-root q.v.) of vertebrates, containing motor fibres.

VENTRICLE. (1) One of the chambers of the heart, the main pumping part, with thick muscular walls. Receives blood from the auricle(s) and passes it to the arteries. Occurs in molluscs and vertebrates. Fish and Amphibia have a single ventricle; reptiles have a partially divided ventricle, except crocodiles where division into two is complete; birds and mammals have two, right supplying blood to lung, left supplying rest of body. (2) Cavity within vertebrate brain filled with cerebrospinal fluid; lateral ventricles in cerebral hemispheres; third in rest of fore-brain; fourth in medulla oblongata.

VENULE. Small vein of vertebrates, differing from capillary by having more connective tissue, and in larger ones smooth muscle, in its wall. Collects blood from capillaries. Permeability of wall of small venule is similar to that of capillary (q.v.).

VERMES. Term formerly used for heterogeneous collection of animals, now recognized to be not closely related. Linnaeus used it to include all animals other than vertebrates and arthropods. Later it was restricted to more or less worm-like animals (Platyhelminthes, Nematoda, Nemertea, Annelida, Rotifera, Polyzoa).

VERNALIZATION. Application of cold treatment to plants to effect flowering. Intimately related with day-length. See *Photoperiodism.* First studied in cereals some of which, if sown in spring, will not flower in the same year but continue to grow vegetatively. Such plants, known as winter varieties, need to be sown in the autumn of the year preceding that in which it is desired that they should flower, and they contrast with spring varieties which, planted in spring, flower in the same year. By vernalization winter cereals can be sown and brought to flower in one season. The seed is moistened sufficiently to allow germination to begin, but not so much as to encourage rapid growth, and when tips of radicles are just emerging, the seed is exposed to a temperature just above 0° C. for a few weeks. Seed thus vernalized acquires properties of seed of spring varieties in that, sown in spring, it produces a crop in the summer of the same year. Vernalization as a practical technique has been developed particularly in Russia. Its significance in agriculture has been demonstrated in that country where it is used, e.g. to avoid killing of winter cereals in the severe winter months. Other possibilities arising from the treatment, e.g. introduction of a crop plant into a new area with a growing season shorter than that under which it normally grows, are being explored.

The cold stimulus is perceived by the apical bud. Successful cold treatment of cereals can be achieved by chilling the isolated embryo rather than the entire seed, while in other plants, e.g. sugar beet, cold treating the apical bud is as effective as treating the entire plant.

VERNATION. Way in which leaves are folded or rolled up in the bud.

VERTEBRA. See *Vertebral Column.*

VERTEBRAL COLUMN (SPINAL COLUMN). Backbone. Longitudinally arranged chain of small bones or cartilages (*vertebrae*) near dorsal side, surrounding spinal cord, characteristic of vertebrates (q.v.). In most vertebrates (not cyclostomes) the vertebrae form a jointed rod, attached to the skull in front, enclosing the spinal cord, and more or less replacing notochord during embryonic development. Each vertebra arises level with the interval between successive somites in the embryo. Each vertebra usually consists of a substantial mass (the centrum) where the notochord was, with an arch (neural arch) above enclosing the spinal cord and often a similar arch (haemal arch) below, or ventral projections, enclosing main axial blood-vessels. They are jointed to each other by their centra, and often by projections of the neural arch, though restricted movement only is possible between any two vertebrae. See *Symphysis*. In fish all vertebrae from skull to tail are very similar. But in tetrapods there are regional differences usually as follows: atlas and axis (q.v.); neck (cervical) with reduced ribs; thoracic, bearing ribs; lumbar, again without ribs; sacral, attached by rudimentary ribs to pelvic girdle; and tail (caudal).

VERTEBRATA (CRANIATA). Most important sub-phylum of the phylum Chordata. Contains the fish, amphibians, reptiles, birds and mammals, i.e. classes Agnatha, Placodermi, Choanichthyes, Actinopterygii, Chondrichthyes, Amphibia, Reptilia, Aves, Mammalia. Vertebrates differ from other chordates in having a skull, which surrounds a well-developed brain, and a skeleton of cartilage or bone. Not all the fossil ones have been shown to have a vertebral column (though most of them have), so that the term Craniata is perhaps preferable, though in much less common use. Vertebrata is sometimes used as synonym for Chordata.

VESICULA SEMINALIS. Seminal vesicle (q.v.).

VESSEL (TRACHEA). Non-living element of xylem (q.v.) consisting of a tube-like series of cells arranged end to end, running parallel with long axis of the organ in which it lies and in communication with adjacent elements by means of numerous pits in side walls. Component cells of a vessel, known as vessel segments, cylindrical in shape, sometimes broader than long, with large perforations in end walls. Functioning in conduction of water and mineral salts, and for mechanical support. Cf. *Tracheid.*

VESTIGIAL ORGAN. Organ, the size, structure, and function of which have diminished and simplified in the course of evolution until only a trace remains. Such vestiges may still have important function in organizing embryonic development, even if their former adult function has disappeared.

VIABLE. Able to live. Of seeds or spores, able to germinate.

VIBRISSAE. Whiskers. Stiff hairs projecting from face, and sometimes limbs, of most mammals, acting as touch receptors.

VICARIAD. One of a group of (vicarious) closely related species whose distribution is allopatric (q.v.).

VILLUS. Finger-like projection. *Intestinal villi* (about 1 mm. long in man), in enormous numbers, give lining of small intestine of many vertebrates velvet-like appearance. Each villus is covered by absorptive epithelium, contains blood-vessels and a lacteal, and moves continually by smooth muscle within it. Absorptive surface of intestine is thus immensely increased. *Chorionic villi*: projections of chorion (q.v.) in mammalian placenta (q.v.) which increase area of contact between embryonic and maternal tissues.

VIRUS. A member of a group of disease-producing agents parasitic in plants and animals, unable to multiply outside the host tissues; invisible by ordinary microscopical methods, so small as to pass through filters which retain bacteria. Of great agricultural and medical importance, causing, e.g. various diseases of potatoes, foot-and-mouth disease, common cold, influenza, smallpox, measles, yellow fever. Plant viruses are usually transmitted by insects, e.g. aphids; some by contact of plants; they are readily transmitted during vegetative propagation, in tubers, bulbs, runners, only occasionally through seed. Animal viruses are spread by insect vectors, by contact, or in droplets of mucus expelled from nose, throat and mouth of infected animal and inhaled by another. Bacteria also suffer from virus attacks. See *Bacteriophage*.

There has been much argument as to whether viruses are living or not. The fact is that they have some of the properties of obviously living things, and not others. Whether they are called living or non-living adds nothing to our knowledge of their properties. Some plant viruses, e.g. tobacco mosaic, consist of nucleic acid and protein and they can be obtained in crystalline form. All grades of complexity exist however among viruses, up to large ones like that of vaccinia (about 0.2 micron diameter), which has a complex internal structure and contains many different substances including various enzymes.

VISCERAL ARCHES. (1) The series of partitions on each side of pharynx between adjacent gill-slits, or between mouth and first gill-slit (spiracle) in fish; or between the corresponding gill-pouches or gill-slits of tetrapod embryos. (2) The skeletal bars (cartilage or bone) lying in these partitions in fish. The bars of each side meet the corresponding bars of the opposite side mid-ventrally, so that together they form a series of inverted arches beneath mouth and throat. The first ('mandibular') arch becomes the primitive jaw skeleton (palatoquadrate and Meckel's cartilage); the second ('hyoid') arch, behind the spiracle, is usually concerned in attachment of jaws to skull (but see *Placodermi*); and the remainder ('branchial arches') lie behind functional gill-slits.

VISCUS. Organ lying in abdominal cavity.

VISUAL PURPLE. Compound of a protein with a prosthetic group of which Vitamin A is an essential precursor. Occurs in rods (q.v.) of vertebrate retina. Chemically changed (and bleached) by light, the change being basis of function of rods as light-sensitive receptors. Increases in amount in dark, raising sensitivity of rods to faint light (dark adaptation).

VITAL STAINING. Staining of cells, or parts of them, while they are living. Several (basic) dyes stain cells placed in them (often called *supravital staining*): e.g. neutral red stains vacuoles and granules in many cells, Janus green stains chondriosomes specifically, methylene blue stains axons. Used in determining morphogenetic movements in embryos. Other (acid) dyes, e.g. trypan blue when injected into an animal (often called *intra vitam staining*) are taken up by cells, especially by the macrophages of vertebrates.

VITAMIN. Organic substance which an organism must obtain from its environment, though it is necessary only in minute amounts. The essential amino-acids (q.v.), which are similarly needed, but in larger amounts, are not included amongst vitamins. A vitamin plays an essential role in metabolism, probably often as part of an enzyme system (hence the small amounts required), and in so doing is broken down, like many other substances in the organism, and lost. What distinguishes a vitamin from most other compounds essential in metabolism is that the organism cannot replace its loss by synthesizing it, and so must obtain it from outside. Occasionally an animal may be able to synthesize part of its requirement (e.g. D by man), and often symbiotic organisms provide part of all of the requirement (e.g. B vitamins synthesized by gut bacteria of insects and vertebrates). Vitamins are required only by heterotrophic organisms; autotrophic organisms, such as most of the green plants, are by definition independent of an external supply of organic compounds. What is a vitamin for one heterotrophic organism may be synthesized in adequate amounts by, and is therefore not a vitamin for, another; or it may even take no part at all in the metabolism of another. There is no such thing as a vitamin in general, but only for specified organisms. There may be mutant strains within one species, some requiring a vitamin, others synthesizing it. Some (like those of the B complex) which are perhaps universal *constituents* of existing organisms are however required by a very wide range of organisms; others (like C) by very few. Every vitamin necessary for a given organism is synthesized by other organisms, otherwise a continuous supply will not be available. The vitamin requirements of the vast majority of organisms are quite unknown. Sometimes several different compounds can substitute for each other in satisfying a given requirement (e.g. see *Vitamin D*); either because the organism requires, not a specific molecule, but a specific chemical grouping which is present in, and available from, each of the alternative compounds; or because conversion of a few closely related groupings into the one required is possible within the organism. Deficiency of the vitamin reduces the rate of the (usually unknown) metabolic process in which it is concerned, with widespread effects (symptoms of deficiency disease). A general effect of deficiency of most vitamins, which was important in the early history of their discovery, is that growth of young animals is stunted.

VITAMIN A. Required by man and other vertebrates but not by other organisms. Fat soluble. Stored in liver. Deficiency in man causes night-blindness (the vitamin being a constituent of visual purple) and keratinization of cornea, and other effects involving epidermis and nervous tissue. Carotene (q.v.) which is synthesized by green plants, and some related pigments, can be converted by animals to Vitamin A, and the vertebrates obtain their supply of the vitamin ultimately from this source.

VITAMIN B COMPLEX. Several water-soluble vitamins, formerly thought to be a single vitamin. See *Thiamin* (B_1), *Riboflavin* (B_2), *Pyridoxine* (B_6), *Pantothenic Acid, Nicotinic Acid, Cobalamine* (B_{12}).

VITAMIN C. Ascorbic acid (q.v.).

VITAMIN D. Required as a vitamin by man, though some is synthesized; and by other vertebrates. Fat soluble. Stored in liver. Deficiency in childhood causes rickets; in adults, sometimes osteomalacia. Effect of the vitamin is on absorption and deposition of calcium and phosphate. Synthesis in man is by action of ultraviolet (in sunlight) on a precursor in skin. Several slightly different compounds have Vitamin D activity for man, though only one has for fowls. All are sterols.

VITAMIN E (TOCOPHEROL). Required as a vitamin by vertebrates. Fat soluble. Deficiency causes abortion, and sterility in male.

VITAMIN F. Linoleic acid (q.v.).

VITAMIN K. Required as a vitamin by man, other mammals, and birds. Fat soluble. Deficiency causes proneness to haemorrhage, since the vitamin takes part in synthesis of prothrombin in liver. Various slightly different compounds have Vitamin K activity. Has been synthesized in water soluble form. Part of human requirements probably come from bacteria living in gut.

VITELLINE MEMBRANE. Membrane which immediately surrounds ovum (q.v.) of animals. It is secreted by the ovum. E.g. membrane enclosing the yolk of a hen's egg.

VITREOUS HUMOUR. Jelly-like material which fills cavity of vertebrate eye behind the lens. See Fig. 3, p. 85.

VIVIPAROUS. Having embryos which develop within the maternal organism and derive nutriment by close contact with maternal tissues, frequently by a placenta, without interposition of any egg-membranes. Viviparity occurs in all placental mammals, and sporadically in other groups of animals. Cf. *Ovoviviparous, Oviparous.* (Bot.). Having seeds which germinate within the fruit, e.g. mangrove, or producing shoots for vegetative reproduction instead of an inflorescence, as in some grasses.

VOCAL CORD. See *Larynx.*

VOLUNTARY MUSCLE. Striped muscle (q.v.).

W

WANDERING CELLS. Cells of blood and connective tissue which actively migrate *in vivo*, e.g. macrophages, amoebocytes.

WARNING (APOSEMATIC) COLORATION. Conspicuous markings on an animal which is poisonous, distasteful, or similarly defended against predators. Coloration is presumed to facilitate learning, on the part of predators, to keep off. See *Mimicry*.

WATER VASCULAR SYSTEM. Of echinoderms, a system of canals containing fluid, open to either the sea or body-cavity, which supplies fluid to tube-feet (q.v.).

WHALEBONE. Baleen (q.v.).

WHEEL ANIMALCULES. Rotifera (q.v.).

WHITE BLOOD CELL. Cell of animal blood, containing no respiratory pigment. In vertebrates, polymorphs, lymphocytes, and monocytes.

WHITE MATTER. Tissue of central nervous system of vertebrates, mainly formed of nerve fibres whose myelin gives it glistening white appearance. Glia and blood-vessels also present. Lies as a layer predominantly outside grey matter (q.v.), forming the tracts connecting different parts of the central nervous system together.

WHOLE MOUNT. An entire organism or part of an organism, prepared by fixation, staining, dehydration, and clearing (but not cut into sections by microtome), and placed on a slide, usually in Canada balsam under a coverslip, for microscopical examination.

WILD TYPE. Phenotype which is characteristic of the majority of individuals of a species in natural conditions. See *Dominance*.

WILTING. Condition of plants in which cells lose their turgidity, and leaves, young stems, etc., droop; resulting from excess of water loss (in transpiration) over water absorption.

WOLFFIAN DUCT. Vertebrate kidney duct, one on each side. Formed in embryo of all vertebrates from region of pronephros and subsequently taken over by mesonephros becoming its duct. In adult anamniotes is usually the kidney duct, and testis also discharges sperm through it. In adult amniotes, where functional kidney (metanephros) has separate duct, it persists only in male, for discharge of sperm, and forms epididymis and vas deferens.

WOOD. Xylem (q.v.).

X

XANTHOPHYLL. Photosynthetic yellow pigment (dihydroxy carotene $C_{40}H_{54}(OH)_2$) occurring in chloroplasts and in plastids in other plant parts where chlorophyll is absent and there associated with carotene in producing yellow and orange colours.

X-CHROMOSOME. A sex-chromosome found paired in the homogametic sex and single in the heterogametic sex. Unlike Y-chromosome (q.v.) contains numerous genes, which show sex-linkage (q.v.).

XENIA. Changes due to effect of foreign pollen in visible characters of endosperm. Result from influence of gene complement of male nucleus which fuses with primary endosperm nucleus as a prelude to endosperm formation. Classical examples afforded by maize, e.g. when variety with white endosperm is pollinated by one with deep yellow endosperm, endosperm of resulting seeds is pale yellow.

XENOPUS. African clawed toad. Used in pregnancy diagnosis, since produces eggs when injected with urine of pregnant woman.

XEROMORPHY. (Bot.). Possession of morphological characters associated with xerophytes.

XEROPHYTE. Plant of dry habitat able to endure conditions of prolonged drought, e.g. in areas of low rainfall such as deserts; due either to capacity during brief rainy seasons for internal storage of water that is used when none can be obtained from soil together with low daytime transpiration rate associated with closure of stomata, e.g. in succulents such as cacti; or to ability to recover from partial desiccation, e.g. desert shrubs such as creosote bush. As long as water is freely available there is evidence that non-succulent xerophytes transpire as, or more freely, than do mesophytes but unlike the latter they can endure periods of permanent wilting during which transpiration is reduced to very low levels. Features associated with this reduction, in different xerophytes, include dieback of leaves which cover and protect perennating buds at soil surface, shedding of leaves when water supply is exhausted, heavily cutinized or waxy leaves coupled with closure or plugging of stomata, sunken or protected stomata, orientation or folding of leaves to reduce insolation, microphyllous habit reducing possibility of drought necrosis of mesophyll. Cf. *Hydrophyte*, *Mesophyte*.

XEROSERE. Type of sere (q.v.).

XYLEM. Wood. Vascular tissue that conducts water and mineral salts throughout the plant and provides it with mechanical support. Of two kinds, *primary*, formed by differentiation from procambium and consisting of protoxylem (q.v.) and metaxylem (q.v.), and *secondary*, additional xylem produced by activity of cambium. Characterized by presence of tracheids and/or vessels, fibres, and parenchyma. In mature woody plants makes up bulk of vascular tissue itself and of entire structure of stems and roots, e.g. tree trunk.

Y

Y-CHROMOSOME. The sex-chromosome (q.v.) found only in the heterogametic sex. Usually differs from the X-chromosome (q.v.) in size. Often only a short part of it pairs with the X-chromosome at meiosis, and it usually contains few or no major genes.

YEASTS. Widely distributed unicellular Fungi, belonging mainly to the Ascomycetes, which multiply typically by a budding process. Of great economic importance. The brewing and baking industries depend upon capacity of yeasts to secrete enzymes that convert sugars into alcohol and carbon dioxide. In the former industry alcohol is the important product; in the latter escaping carbon dioxide causes dough to rise. Yeasts are also used commercially as a source of proteins and vitamins.

YOLK. Store of food material, in form of protein and fat granules, in eggs of majority of animals.

YOLK-SAC. Sac containing yolk which hangs from ventral surface of very yolky vertebrate embryos (elasmobranchs, teleosts, reptiles, birds). Has outer layer of ectoderm; inner layer, usually absorptive, of endoderm; with mesoderm containing coelom and blood-vessels between. The endoderm is in effect a diverticulum of the gut, the yolk- or fluid-filled space within it communicating with intestine of embryo. In mammals the yolk-sac, quite empty of yolk, is only the endodermal sac, with its layer of mesoderm, homologous with that of the other vertebrates; the ectodermal layer with its mesoderm is called part of the chorion. The same terminology is often applied to the other amniotes. As absorption of yolk proceeds, yolk-sac is withdrawn until it merges into embryo; but most of it is cut off from the embryo at birth in mammal. See Fig. 8, p. 184.

Z

ZONA PELLUCIDA. Homogeneous (mucoprotein) membrane around the mammalian egg, which disappears before implantation in the uterus.

ZOOID. Member of a colony of animals which are joined together, e.g. polyp, polypide.

ZOOLOGY. Study of animals.

ZOOPLANKTON. Animals of plankton (q.v.). Cf. *Phytoplankton.*

ZOOSPORANGIUM. (Bot.). Sporangium producing zoospores; present in certain Fungi and Algae.

ZOOSPORE (SWARM-SPORE). Naked spore produced within a sporangium (zoosporangium), motile by one, two, or many flagella;

present in certain members of phycomycete Fungi and green and brown Algae.

ZYGOMORPHIC. See *Bilateral Symmetry.*

ZYGOSPORE. Thick-walled resting spore, product of sexual reproduction by fusion of contents of two similar gametangia; found in a group of phycomycete Fungi, Zygomycetes, and in an order of green Algae, Conjugales.

ZYGOTE. The fertilized ovum, before it undergoes cleavage.

ZYMASE. Enzyme system of yeast which breaks down (ferments) hexose sugars into alcohol and carbon dioxide. The first cell-free enzyme preparation (1897).